An introduction to the principles of surface chemistry

Cambridge Chemistry Texts

GENERAL EDITORS

E. A. V. Ebsworth, Ph.D.
Professor of Inorganic Chemistry
University of Edinburgh

D. T. Elmore, Ph.D.
Professor of Biochemistry
The Queen's University of Belfast

P. J. Padley, Ph.D.
Lecturer in Physical Chemistry
University College of Swansea

K. Schofield, D.Sc.
Reader in Organic Chemistry
University of Exeter

An introduction to the principles of surface chemistry

R. AVEYARD
Lecturer in Chemistry, University of Hull

AND

D. A. HAYDON
Assistant Director of Research, Department of Physiology
University of Cambridge, and Fellow and Director of Studies
in Natural Sciences, Trinity Hall

CAMBRIDGE
at the University Press 1973

Published by the Syndics of the Cambridge University Press
Bentley House, 200 Euston Road, London NW1 2DB
American Branch: 32 East 57th Street, New York, N.Y.10022

Library of Congress Catalogue Card Number: 72–89802

ISBNs
0 521 20110 1 hard covers
0 521 09794 0 paperback

Printed in Great Britain
at the University Printing House, Cambridge
(Brooke Crutchley, University Printer)

To our chemistry teachers

Contents

Preface

This book is intended to be an introduction to the physical chemistry of surfaces. Some knowledge of thermodynamics, statistical thermodynamics and electrochemistry is assumed and, for this reason, the level is appropriate for final year undergraduates or graduates. The emphasis throughout the book is on the basic principles of the subject rather than on the presentation of factual information, and a special attempt has been made to highlight the common theoretical background to the treatments of the various types of interface. Experimental techniques other than those employed for the determination of adsorption at liquid and at solid surfaces have not been described in any detail. However, references which cover practical methods have been given.

Many of the fundamental concepts of the subject are introduced in chapters 1 and 2. Thus, chapter 1 is concerned mainly with the thermodynamics and statistical thermodynamics of surfaces, while chapter 2 is devoted to electrical phenomena. In subsequent chapters specific types of interface are discussed. Liquid interfaces, which are relatively simple, are described in chapter 3. Electrode surfaces, particularly mercury–aqueous solution and silver-halide–aqueous solution interfaces, are dealt with in chapter 4, while chapters 5 and 6 are concerned with solid surfaces in contact with, respectively, gases and liquids. It is not essential to read the chapters in the order that they appear. Although chapters 1 and 2 provide the theoretical basis for much of what follows, they may best be approached through the back references in later chapters. In order to facilitate a first reading, the more important equations have been highlighted, e.g. ◀ **(1.8)**.

S.I. units have been used throughout the book but, for convenience, the more commonly used equivalent has frequently been quoted. In addition a short conversion table is included. Since the S.I. is a rationalised system, some of the electrical equations may have an unfamiliar appearance, but this, of course, arises only from the omission of 4π.

We are indebted to many friends and colleagues for advice and assistance. Drs J. N. Agar and R. Scowen kindly read sections of the manuscript and offered valuable criticisms. Less directly involved are those who have already pioneered the description of surface chemistry. Inevitably, in a book of this type, much of the material is drawn from existing treatments of the subject and, in some instances, we have thought it undesirable to change appreciably a previous approach. In these instances our debt to earlier writers is considerable and is, we hope, fully acknowledged. On questions of presentation Dr P. J. Padley has, from the outset, provided much helpful guidance, and we wish to thank him for his patience in even the smallest matters. Finally we thank our wives for their forebearance and especially Primrose Haydon for preparing the typescript.

Cambridge R. A.
June 1972 D. A. H.

Symbols

The more important symbols used in the text are listed below. Where it is useful, the number of the equation or section in which the symbol is introduced is given in brackets.

A	Helmholtz free energy
$\mathcal{A}$	interfacial area
a	partial molar surface area
$\dot{a}$	area per molecule at the surface
$\dot{a}_0$	co-area of a molecule at the surface
$\boldsymbol{a}$	activity
$\boldsymbol{a}_{\pm}$	mean ionic activity (3.85)
C	capacitance
c	number of nearest neighbour sites per site (1.97)
c	number of components (3.50)
c	constant in BET equation (5.37)
E	electromotive force
$\boldsymbol{E}$	electric field strength
e^-	charge of electron
G	Gibbs free energy (1.18)
$\mathcal{G}$	Gibbs free energy (1.17)
g	gravitational acceleration
H	enthalpy (§5.22)
$\mathcal{H}$	enthalpy (§5.22)
h	Planck's constant
$j(T)$	internal partition function of a molecule (§1.10)
k	Boltzmann's constant
$l(T)$	translational partition function of a molecule
m	mass
N	number of molecules
N_A	Avogadro's number
n	number of moles
p	pressure
p^0	normal vapour pressure

p^s normal component of surface dipole moment per molecule (2.20)
Q heat adsorbed by system
q differential heat of adsorption (5.73)
R gas constant
S entropy
T absolute temperature
U internal energy
V potential energy
V volume
V_m molar volume
V_i partial molar volume of i
ΔV surface Volta potential difference
v velocity (chapter 2)
v volume of gas adsorbed (chapter 5)
v_m monolayer capacity
W work done on system
W_A work of adhesion (3.20)
W_C work of cohesion (3.19)
x distance in x direction (2.23)
x mole fraction
x relative pressure
Z partition function of system
z charge number of an ion
$\mathfrak{z}(T)$ complete partition function of a molecule (1.52; 1.80)
α two-dimensional van der Waals constant (1.77)
α coefficient of accommodation (5.1)
α_i^α real potential of i in phase α (2.6)
Γ surface concentration or surface excess (1.40)
γ surface tension
Δ change in a quantity
Δ_a change in a quantity accompanying adsorption
δ effective ionic radius
ϵ adsorption potential (§5.13)
ϵ_r dielectric constant
ϵ_0 permittivity of free space
ζ electrokinetic or zeta potential
ζ $(\partial A/\partial n_i)_{T,V,n_j,\gamma}$
η viscosity
θ surface coverage
θ contact angle

κ reciprocal length parameter (2.33)
μ chemical potential
$\tilde{\mu}$ electrochemical potential
π surface pressure
π mathematical constant
ρ density
ρ space charge density (2.23)
Σ specific surface area of a solid
σ surface charge density (2.27)
σ spreading tension (spreading coefficient) (3.23)
τ residence time of molecule at a surface (5.2)
Φ interaction energy
ϕ Harkins–Brown correction factor (3.14)
ϕ volume fraction (3.43)
φ number of bulk phases (3.50)
φ inner electrical potential
φ volume of adsorbed material at a surface (§5.13)
χ surface or chi-potential
ψ outer electrical potential
ψ number of surface phases (3.50)
ω number of degrees of freedom (3.50)

Subscripts

i chemical component i
σ specific surface excess

Superscripts

g gas phase
l liquid phase
0 pure substance
s total surface quantity
α phase alpha
β phase beta
σ general index for surface; excess surface quantity (see §1.7)
$\ominus$ standard state

Units of commonly occurring quantities

Quantity	Previously used unit	S.I. unit (or sub-multiple)	Factor†
Area	In surface chemistry the use of $Å^2$ is very common	nm^2	10^{-2}
Energy	erg	J	10^{-7}
	calorie		4.184
Pressure	atmosphere	$N\ m^{-2}$	1.01325×10^5
	dyne cm^{-2}		10^{-1}
	mmHg		1.33322×10^2
Surface tension Surface pressure	dyne cm^{-1}	$mN\ m^{-1}$	1
Surface free energy etc.	erg cm^{-2}	$mJ\ m^{-2}$	1
Electric charge	esu	C	3.335×10^{-10}
	emu		10
Capacitance	$esu^2\ erg^{-1}$	F	1.112×10^{-12}

† Value in S.I. unit (or sub-multiple) = value in previously used unit × factor.

1 Some general principles relating to surfaces

1.1. Introduction. The boundary region between two adjacent bulk phases is known as an *interface* although, when one of the phases is a gas or a vapour, the term *surface* is commonly used.

An interface may be considered, from a thermodynamic standpoint, either as a mathematical plane or as a distinct phase having a finite thickness. For the purposes of constructing a molecular model the interface must be at least one molecular diameter in thickness, and, in rigorous theoretical treatments, the interface may be considered more realistically as extending over several molecular thicknesses. Matter at an interface usually has different physical properties and energy characteristics from that in the bulk, and the study of interfaces has grown into a distinct branch of chemistry – surface chemistry.

It is convenient to treat different types of interface separately. This is mainly because of the differences in physical characteristics exhibited by solid and liquid surfaces and the ensuing necessity for the use of totally different experimental techniques. The usual classification is as follows:

Interfaces

- LIQUID–LIQUID (L/L)
  - LIQUID–VAPOUR (L/V)
- SOLID–GAS (S/G)
- SOLID–LIQUID (S/L)

The letters in brackets are the abbreviations sometimes employed.

In later chapters it will be necessary to deal with the three types of interface separately but, in this chapter, general relationships and properties which, at least in principle, apply to any interface will be described. In this way it is hoped to underline the basic similarities between the various types of interface. These similarities are often obscured partly because of the different experimental procedures involved in their study and partly because of the historical development of the subject.

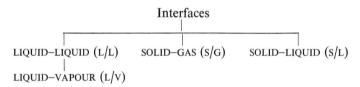

1.2. Surface tension and surface free energy. Consider a drop of a one component liquid in equilibrium with its vapour and in the absence of any external forces. The drop spontaneously assumes the form of a sphere, the shape which corresponds to the minimum surface to volume ratio. It can therefore be inferred that work must be done on the drop to increase its surface area, and hence that the surface molecules are in a state of higher free energy than those in the bulk liquid. This is in part explained by the fact that surface molecules have fewer nearest neighbours and, as a consequence, fewer intermolecular interactions than bulk molecules. There is then, a free energy change associated with the isothermal, reversible formation of a liquid surface and this is termed the *surface free energy* or, more correctly, the *excess surface free energy.* The terms are often used to mean specific (i.e. per unit area) surface free energy, the units of which are mJ m^{-2}. It must be emphasised that this surface free energy is not the *total* free energy of the surface molecules but the excess free energy which the molecules possess by virtue of their being in the surface (see §1.6 and §1.7). It is sometimes convenient to consider the total surface free energy, however, and this will be introduced in §1.6 and §1.7.

It is easily understood from the above observations that surface molecules are subject to an inward attraction normal to the surface. This is equivalent to saying that the surface itself is in a state of lateral tension and leads to the concept of *surface tension.* For a plane surface, the surface tension is defined as the force acting parallel to the surface and at right angles to a line of unit length anywhere in the surface. For a curved surface the definition, while similar, is slightly more complex. Nevertheless, the magnitude of the surface tension is not a function of the curvature except for very small radii of curvature (see §3.2). The units of surface or interfacial tension are mN m^{-1} and the symbol used here will be γ, although σ and ρ are commonly employed by other authors. The units of surface tension and of specific excess surface free energy are dimensionally equivalent (MT^{-2}) and, for pure liquids in equilibrium with the vapour, the two quantities are numerically equal. Both are intensive thermodynamic properties, whereas the surface excess free energy is an extensive property and is dependent on the interfacial area of the system. The explicit thermodynamic definition of surface tension for a pure liquid is

$$\gamma = \left(\frac{\partial A}{\partial \mathcal{a}}\right)_{T,V}$$

where A is the Helmholtz free energy of the system and $\mathscr{A}$ is the area of the surface. This definition will be discussed in more detail in §1.7. It must be stressed that surface tension and specific excess surface free energy are numerically equal only in systems where no adsorption occurs (§1.7).

Consider now a system of two pure immiscible liquids which make contact at a plane interface. The forces acting on the surface molecules are similar to those encountered in the liquid–vapour system but, on replacement of the vapour by a condensed phase, the mutual attraction of unlike molecules across the interface becomes important. The free energy required to form fresh interface is referred to as the *excess interfacial free energy*; the specific excess interfacial free energy for pure liquids is dimensionally equivalent and numerically equal to the *interfacial tension*.

So far no reference has been made to solid surfaces. In principle, all that has been said in connection with liquid interfaces applies. That is, the specific excess surface free energy of a pure solid is the work required to form unit area of surface, and this is equivalent to the surface tension of an equilibrium isotropic surface. In practice, however, the position is rather different. When a fresh solid surface is formed by cleaving the solid in a vacuum or, if the solid is volatile, in its own vapour, the ions or molecules in the freshly generated surface will normally be unable (owing to their immobility) to take up their equilibrium configuration. The surface so formed is therefore a non-equilibrium structure. This situation is quite distinct from that for a liquid surface which, since the molecules are in a fluid state, attains equilibrium almost as soon as it is formed. It is therefore convenient in the case of solids to define surface tension in terms of the restoring force necessary to bring the freshly exposed surface to its equilibrium state (see §5.4).

The units of surface tension, $mN\,m^{-1}$, are the two-dimensional analogues of the bulk pressure units, $mN\,m^{-2}$. Surface tension may be regarded, therefore, as a two-dimensional negative pressure.

It will be seen later that the surface tension of liquids can be measured simply and directly, whereas the measurement of the surface tension and other thermodynamic parameters for solid surfaces is a difficult and often ambiguous procedure.

1.3. Adsorption and surface pressure. For present purposes adsorption may be considered to be the partitioning of a chemical species between a bulk phase and an interface. Thus, when a solid is in equilibrium with a gas, the gas is usually more concentrated in the region of the surface. In this instance the gas is said to be *positively adsorbed.* There are systems, however, where a species is less concentrated in the interfacial region than in the bulk, and then the substance is said to be *negatively adsorbed.* The latter state of affairs exists, for example, in aqueous solutions of some simple inorganic electrolytes, where it is found that there is a surface deficiency of electrolyte. The factors responsible for adsorption will be discussed in subsequent chapters.

Adsorption is quite distinct from absorption, but the two processes may often take place simultaneously. If, for instance, ethanol vapour were allowed to attain equilibrium with water, the ethanol would in part dissolve in the water, but it would also adsorb positively at the water–vapour surface. The general term which includes both adsorption and absorption is simply *sorption.*

When a system consists of more than one component the surface tension (defined now as $(\partial A/\partial \mathcal{A})_{T,V,n_i}$ where n_i means that all the numbers of moles $n_1 \ldots n_i$ of the various components i are kept constant) may vary with the composition. The variation of surface tension with composition is formally described by the Gibbs adsorption equation (§1.8) which shows that positive adsorption always leads to a lowering of γ. If γ_0 denotes the value when no adsorption has occurred and γ when adsorption has occurred, then the *surface pressure* or *speading pressure*, π, is defined as $\pi = \gamma_0 - \gamma$. It can be considered as a two-dimensional pressure exerted by the adsorbed molecules in the plane of the surface.

The thickness of adsorbed layers depends very much on the nature of the system, but usually adsorption at liquid–vapour and liquid–liquid interfaces yields films one molecule in thickness, i.e. *monolayers.* In solid–gas adsorption, monolayers can result if the adsorbate in the bulk is at a sufficiently low pressure, but at higher pressures films of several molecular layers, known as *multilayers*, tend to build up.

The nature of the forces between the adsorbed molecules and the surface is the main factor in determining the type of adsorption process. Thus, forces may be of the van der Waals type or they may arise from the formation of chemical bonds between the adsorbed molecules and the surface. In the former event the adsorption is called *physical adsorption* and in the latter, chemical adsorption, or more usually, *chemisorption.*

The most striking difference between the two types of adsorption is the magnitude of the heat of adsorption. Since physical adsorption involves only van der Waals forces, the adsorption heats are small, ranging usually between 0 and 20 kJ per mole of adsorbate. Heats of chemisorption, however, which reflect chemical bond formation, are much higher and values ranging from 80 to 400 kJ mol^{-1} are common. Many gases are chemisorbed on to the surfaces of metals and metal oxides; the adsorption of ethylene on to zinc oxide, for example, gives a heat of 100 kJ mol^{-1} and hydrogen on to nickel between 80 and 120 kJ mol^{-1}, depending on the state of the nickel. Adsorption at liquid–liquid and liquid–vapour interfaces is generally physical in nature, but very many instances are known of physical adsorption at gas–solid interfaces.

There are also other differences between the two types of adsorption. Physically adsorbed molecules may easily be removed or desorbed from the surface by lowering the bulk concentration or pressure of the adsorbate. For example, if an aqueous solution of ethanol is diluted with water, the surface tension will rapidly assume a higher equilibrium value indicating that the adsorbed molecules will readily desorb. In addition, this solution will have the same surface tension as if it had been formed from a more dilute solution by addition of pure ethanol. Similarly, a physically adsorbed gas will usually leave the surface of a solid quite rapidly if the pressure of the gas is reduced. (Often, however, the amount adsorbed at a given pressure depends on whether this pressure is reached from a higher or a lower value. This phenomenon is termed hysteresis and is dealt with in §5.19.) Chemisorbed species on the other hand are much more difficult to remove, and high temperatures and very low pressures are often required to bring about desorption. Chemisorbed films are usually monolayers, whereas physically adsorbed films on solids are frequently multilayers.

This book is mainly concerned with physical adsorption; readers interested in chemisorption are recommended to consult Hayward & Trapnell (1964).

1.4. Types of surface. The heat of adsorption of a species from a dilute vapour on to a surface is made up of two contributions: (*a*) the heat of interaction of the adsorbate molecules with the surface and (*b*) the heat which arises from the lateral interactions between adsorbate molecules. If it is assumed that the films are sufficiently dilute for these lateral interactions to be absent, and that only the adsorbate–adsorbent interactions are important, a distinction between homogeneous and heterogeneous surfaces may readily be made.

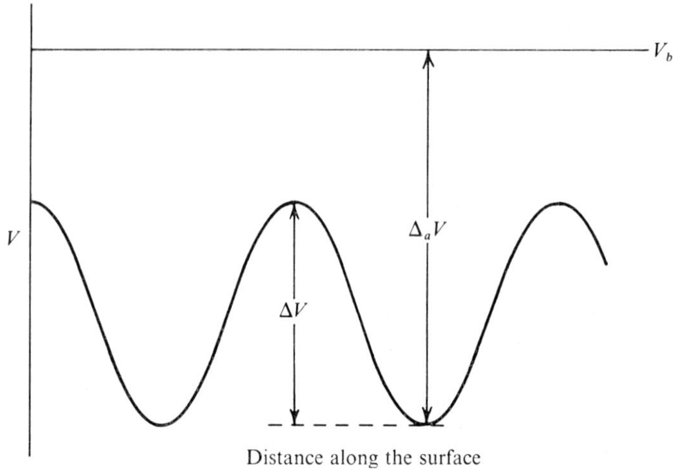

Fig. 1.1. Variation of potential energy with position along the surface.

A homogeneous surface is one which is energetically uniform on a submicroscopic (but not molecular) level and the heat of adsorption is independent of the fraction of surface covered. Liquid surfaces are homogeneous since the surface molecules are fluid and any non-uniformity is very short-lived. The absence of fluidity in solids gives rise often to heterogeneous surfaces, i.e. those which are energetically non-uniform and, consequently, heats of adsorption are found which vary with the area of surface covered. It is possible to prepare solids with more or less homogeneous surfaces, but often the presence of the slightest heterogeneity can cause serious changes in the adsorption characteristics, especially for very dilute surface films.

1.5. Localised and non-localised films. On a molecular level the potential energy of interaction of an adsorbate molecule with a surface varies with its position on the surface because of the adsorbent structure, as illustrated in fig. 1.1. Here V_b is the potential energy of the adsorbate molecule in the bulk phase and $\Delta_a V$ is the difference between V_b and the minimum potential energy for the molecule on the surface. As a consequence of its thermal energy the adsorbed molecule vibrates around the minimum of the potential well, of depth ΔV. The average position of the molecule during the vibration is called an *adsorption site*.

If ΔV is much greater than the thermal energy, the adsorbed molecules spend most of their time on the surface in or close to the minima of the potential energy wells and are said to be *localised.* When, on the other hand, ΔV is less than or comparable to the thermal energy, the molecules may readily move along the surface from one site to another, and are said to be *non-localised.* It follows that in a non-localised film desorption can take place from any position on the surface, whereas for a localised film desorption occurs from an adsorption site. According to the model of a localised film, molecules may change their positions on the surface but they must do so by desorption and subsequent adsorption on different sites. The term *mobile* has often been used to mean non-localised. In this book mobile is used in the more practical sense to imply that the molecules can change position on the surface. With this terminology, therefore, it is very likely that a localised film will be mobile, since the time between adsorption and desorption of a molecule at a given site is well below that usually required for an experiment.

For the physical adsorption of gases on solid surfaces, ΔV is of the order of 2 kJ mol^{-1} and, at all but the lowest temperatures, the adsorbate is only partially localised. Hill (1946) has pointed out that the transition between localised and non-localised states starts at a temperature around $\Delta V/10k$ and continues to about $\Delta V/k$, and, that within these limits, the film is in neither one state nor the other. Chemisorbed films of gases on solids are normally localised, owing to the high values of ΔV.

In practice it is often difficult to be sure whether a system is non-localised, localised or intermediate between the two. For strong adsorption from dilute solution to liquid–liquid or liquid–vapour interfaces it seems likely, from the foregoing considerations, that the films would be non-localised. In this instance, however, the argument is somewhat artificial as the mobility of the substrate molecules is a complicating factor. Other examples will be encountered where, for reasons of theoretical simplicity, the system is treated as being localised (e.g. liquid mixtures (§3.6) and multilayers in gas–solid adsorption §5.12) but the apparent success of such treatments does not necessarily mean that the system is, in fact, localised.

1.6. Conventions for the thermodynamic treatment of a surface. In real systems, there is a finite distance across an interface in which the properties gradually change from those of one adjacent bulk phase to those of the other. Accordingly, one way of treating a surface is to consider it as a phase which is separate from the adjacent bulk phases, and which has a finite thickness and volume (see e.g. Guggenheim, 1967*a*).

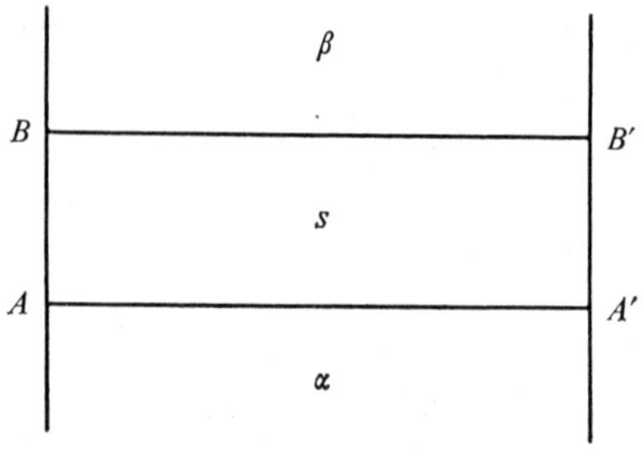

Fig. 1.2. The surface phase.

This phase can be treated thermodynamically in a way somewhat analogous to bulk phases, except that terms relating to γ and interfacial area appear in the thermodynamic expressions. Fig. 1.2 depicts such a model. The regions α and β are homogeneous bulk phases separated by the planar surface phase, *s*. Phase α is homogeneous up to the plane AA' and phase β up to the plane BB'. All changes in properties from α to β take place in the region between AA' and BB'. The surface phase has an arbitrary thickness of not less than one molecular diameter. By the use of this approach, the extensive thermodynamic functions and numbers of moles which appear in thermodynamic expressions relating to the surface phase, are *total* quantities and, where appropriate, will be denoted by a superscript *s*.

The method which Gibbs used to formulate the thermodynamics of surfaces was somewhat different (Gibbs, 1928). In the Gibbs treatment the interface is regarded as a mathematical dividing plane (the *Gibbs surface*). This is illustrated in fig. 1.3 where α and β are homogeneous up to the planes AA' and BB' respectively. The dividing surface is designated SS' and has, of course, zero thickness and volume, and is placed between, and parallel to, AA' and BB' in some arbitrary position. The amount of adsorption of component *i* is measured by its *surface excess*, defined as the amount of *i* in unit area of the region between AA' and BB' less the amount that there would be in the same region if α and β extended unchanged to SS'. In other words, the surface excess is the extra amount of a component in between AA' and BB' by virtue of the presence of the interface. Extensive thermodynamic properties of the surface are defined in an analogous fashion. Parameters defined in terms of the Gibbs surface will be denoted by a superscript σ.

The choice of a particular model for the surface is purely a matter of

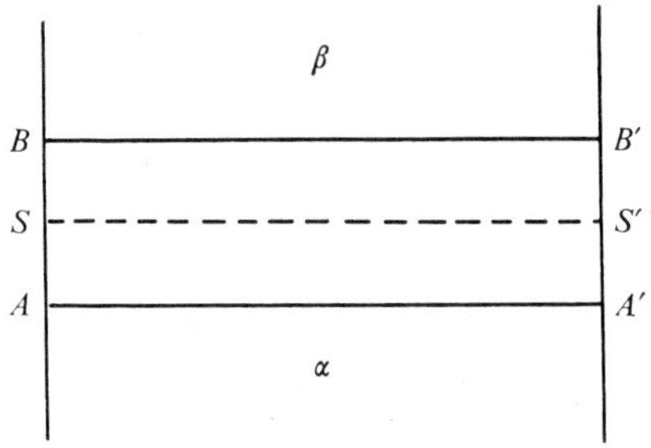

Fig. 1.3. The Gibbs dividing surface.

convenience, both models having advantages and disadvantages, as will be seen later. However, it is sometimes felt that the first approach is conceptually simpler in that it employs a more obvious physical picture of a surface.

1.7. Surface thermodynamic quantities. *Internal energy.* The systems to be considered will consist of two bulk phases α and β and their mutual interface, and it will be supposed throughout that α, β and the interface are in equilibrium with each other.

For a closed system, neglecting surface effects for the moment, the differential expression for the internal energy U of the system is

$$\mathrm{d}U = \mathrm{d}Q + \mathrm{d}W, \tag{1.1}$$

where $\mathrm{d}Q$ is the heat taken up by the system for an infinitesimal change, and $\mathrm{d}W$ is the work done on the system. For a system at equilibrium, infinitesimal changes are reversible and, if the work done is associated entirely with volume changes, $\mathrm{d}W = -p\,\mathrm{d}V$, where V is the volume and p the pressure of the system. Further, $\mathrm{d}Q = T\mathrm{d}S$ where S is the entropy and T the temperature of the system. Equation (1.1) may therefore be written

$$\mathrm{d}U = T\mathrm{d}S - p\mathrm{d}V$$

or, if the system is open

$$\mathrm{d}U = T\mathrm{d}S - p\mathrm{d}V + \sum_i \mu_i \mathrm{d}n_i. \tag{1.2}$$

Here, μ_i is the chemical potential of component i and n_i the number of moles of i in the system; $\sum_i$ denotes the summation over all species $1, \ldots, i$.

If the system contains an interface the work done may, as mentioned in §1.2, change the area of the interface as well as the volume of the adjacent bulk phases. Thus, in this instance

$$dW = -p\,dV + \gamma\,d\mathcal{A},$$

where γ is the tension and $\mathcal{A}$ the area of the interface. Provided the interface is in equilibrium with the bulk, i.e. μ_i is the same in both regions, (1.2) then becomes

$$dU = T\,dS - p\,dV + \gamma\,d\mathcal{A} + \sum_i \mu_i\,dn_i. \tag{1.3}$$

One definition of surface or interfacial tension is therefore

$$\gamma = \left(\frac{\partial U}{\partial \mathcal{A}}\right)_{S,V,n_i}. \tag{1.4}$$

In the discussion that follows it is necessary to choose a notation for surface parameters. As mentioned in §1.6 either superscript s or superscript σ is used, according to the choice of convention for the surface. Up to (1.24), however, the equations presented are valid for either convention and superscript σ has arbitrarily been selected to denote surface properties. Thereafter it will frequently be necessary to specify the convention adopted and both s and σ will be encountered.

The total internal energy of the system comprising α, β and interface is the sum of the internal energies of each of the phases, and so

$$U = U^\alpha + U^\beta + U^\sigma, \tag{1.5}$$

where U^α, U^β and U^σ are the internal energies of α, β and the interface respectively. Correspondingly

$$dU = dU^\alpha + dU^\beta + dU^\sigma. \tag{1.6}$$

There are, of course, similar relationships for other extensive variables.

Equation (1.6) may be expressed in the fuller form

$$dU = \sum_a T\,dS^a - \sum_a p\,dV^a + \gamma\,d\mathcal{A} + \sum_a \sum_i \mu_i\,dn_i^a, \tag{1.7}$$

where $\sum_a$ denotes summation over all the phases present in the system. For the surface alone,

$$dU^\sigma = T\,dS^\sigma - p\,dV^\sigma + \gamma\,d\mathcal{A} + \sum_i \mu_i\,dn_i^\sigma. \qquad \blacktriangleleft\textbf{(1.8)}$$

It should be remembered that when the Gibbs convention for the surface is used $V^\sigma = 0$. No superscript σ has been included for the intensive properties (T, p and μ_i) in (1.7) and (1.8) since at equilibrium these quantities are the same throughout the system.

Helmholtz free energy. There is a choice of useful definitions of the Gibbs and Helmholtz free energy functions for the surface. Consider first the Helmholtz function, for which there are two possibilities:

$$A^\sigma = U^\sigma - TS^\sigma \qquad ◀\textbf{(1.9)}$$

and

$$A^{\sigma,*} = U^\sigma - TS^\sigma - \gamma\mathcal{A} = A^\sigma - \gamma\mathcal{A}. \qquad (1.10)$$

The first of these is the more common definition and is the only one that will be discussed further. The complete differential of A^σ is

$$\mathrm{d}A^\sigma = \mathrm{d}U^\sigma - T\mathrm{d}S^\sigma - S^\sigma\mathrm{d}T. \qquad (1.11)$$

The combination of (1.8) and (1.11) gives

$$\mathrm{d}A^\sigma = -S^\sigma\mathrm{d}T - p\mathrm{d}V^\sigma + \gamma\mathrm{d}\mathcal{A} + \sum_i \mu_i \mathrm{d}n_i^\sigma. \qquad ◀\textbf{(1.12)}$$

This equation refers of course to the surface phase only. For the whole system the Helmholtz free energy is A and

$$\mathrm{d}A = -S\mathrm{d}T - p\mathrm{d}V + \gamma\mathrm{d}\mathcal{A} + \sum_i \mu_i \mathrm{d}n_i. \qquad (1.13)$$

A useful equation is obtained by integration of (1.12) holding constant the intensive properties T, p, μ_i and γ, i.e.

$$A^\sigma = -pV^\sigma + \gamma\mathcal{A} + \sum_i \mu_i n_i^\sigma. \qquad ◀\textbf{(1.14)}$$

From (1.12) and (1.13) two further definitions of surface tension become apparent, namely,

$$\gamma = \left(\frac{\partial A^\sigma}{\partial \mathcal{A}}\right)_{T,V^\sigma,n_i^\sigma} \qquad ◀\textbf{(1.15)}$$

and

$$\gamma = \left(\frac{\partial A}{\partial \mathcal{A}}\right)_{T,V,n_i}. \qquad ◀\textbf{(1.16)}$$

The second of these is the most commonly encountered definition of γ, although (1.15) is particularly useful for systems which contain insoluble monolayers (see §3.7). For a one component system, $\gamma = (\partial A/\partial \mathcal{A})_{T,V}$ as in §1.2, where it is understood that the mass of the system is held constant during the extension of the surface. γ is the isothermal reversible work done in extending the interface by unit area at constant V and n_i.

Gibbs free energy. There are two common and widely used definitions for the Gibbs free energy of the surface, which are analogous to the definitions of the Helmholtz functions. These are

$$\mathscr{G}^\sigma = A^\sigma + pV^\sigma = U^\sigma - TS^\sigma + pV^\sigma \qquad \blacktriangleleft(1.17)\dagger$$

and

$$G^\sigma = A^{\sigma,*} + pV^\sigma = A^\sigma - \gamma\mathscr{A} + pV^\sigma. \qquad \blacktriangleleft(1.18)$$

The differential of (1.17) combined with (1.8) yields

$$\mathrm{d}\mathscr{G}^\sigma = -S^\sigma \mathrm{d}T + V^\sigma \mathrm{d}p + \gamma \mathrm{d}\mathscr{A} + \sum_i \mu_i \mathrm{d}n_i^\sigma \qquad \blacktriangleleft(1.19)$$

or, for the whole system

$$\mathrm{d}\mathscr{G} = -S\mathrm{d}T + V\mathrm{d}p + \gamma \mathrm{d}\mathscr{A} + \sum_i \mu_i \mathrm{d}n_i, \qquad (1.20)$$

where

$$\mathscr{G} = G^\alpha + G^\beta + \mathscr{G}^\sigma.$$

From (1.20) the definition of γ in terms of $\mathscr{G}$ is

$$\gamma = \left(\frac{\partial \mathscr{G}}{\partial \mathscr{A}}\right)_{T,p,n_i}. \qquad \blacktriangleleft(1.21)$$

From the alternative definition of the Gibbs free energy for the surface, given in (1.18)

$$\mathrm{d}G^\sigma = -S^\sigma \mathrm{d}T + V^\sigma \mathrm{d}p - \mathscr{A}\mathrm{d}\gamma + \sum_i \mu_i \mathrm{d}n_i^\sigma. \qquad \blacktriangleleft(1.22)$$

The Gibbs free energy for the system is now written G where

$$G = G^\alpha + G^\beta + G^\sigma$$

and

$$\mathrm{d}G = -S\mathrm{d}T + V\mathrm{d}p - \mathscr{A}\mathrm{d}\gamma + \sum_i \mu_i \mathrm{d}n_i. \qquad (1.23)$$

From the integration of (1.22), holding constant the intensive variables an equation similar to (1.14) is obtained, viz.

$$G^\sigma = \sum_i \mu_i n_i^\sigma \qquad \blacktriangleleft(1.24)$$

which is precisely analogous to the bulk phase equation $G = \sum_i \mu_i n_i$.

As already mentioned, most authors use the definition of the Helmholtz free energy given in (1.9). Both definitions of the Gibbs function (1.17) and (1.18) are used but it is not always stated explicitly which definition is being employed. Table 1.1 shows the thermodynamic parameters used by some authors. Also included is their choice of model for the surface.

† If the Gibbs convention for the surface is employed, $V^\sigma = 0$ so that $\mathscr{G}^\sigma = A^\sigma$.

TABLE 1.1. *Thermodynamic parameters and models for the surface used by some authors*

Function or model	Authors: G	K & O	L & R	H & S	B	A & F
Helmholtz function	A^σ	A^σ	A^σ	A^σ	—	A^σ
Gibbs function	G^σ	G^σ	$\mathcal{G}^\sigma$	$\mathcal{G}^\sigma$	$\mathcal{G}^\sigma$	G^σ
Gibbs dividing surface	—	√	√	—	√	—
Surface phase	√	—	—	√	—	—

G: Guggenheim (1967*b*); K & O: Kirkwood & Oppenheim (1961); L & R: Lewis & Randall (1961); H & S: Hildebrand & Scott (1950); B: Butler (1951); A & F: Aston & Fritz (1959).

Surface tension and surface free energy. It was mentioned in §1.2 that for a pure liquid, γ is a specific excess surface free energy, but that when there is more than one component present and adsorption occurs, this is no longer true. This can be shown in the following way. For the Gibbs model for the surface, as $V^\sigma = 0$, (1.14) becomes

$$A^\sigma = \gamma\mathcal{A}+\sum_i \mu_i n_i^\sigma, \tag{1.25}$$

which, when divided by $\mathcal{A}$, gives

$$A_\sigma = \gamma+\sum_i \mu_i \Gamma_i^\sigma, \tag{1.26}$$

where σ written as a subscript denotes unit area, and Γ_i^σ are surface excess concentrations. Unless $\sum_i \mu_i \Gamma_i^\sigma$ is zero, which would normally occur only if there were no adsorption, the specific excess surface free energy (A_σ) is not equal to γ.

If the alternative 'surface phase' model (fig. 1.2) is used V^s is not zero, but pV^s in (1.14) is very small compared with the other terms and may usually be neglected. For unit area (1.14) then becomes

$$A_s = \gamma+\sum_i \mu_i \Gamma_i^s,$$

where A_s is the specific total surface free energy and Γ_i^s are total surface concentrations. $\sum_i \mu_i \Gamma_i^s$ is never zero and even for a pure liquid A_s does not equal γ.

Chemical potentials. No treatment has so far been given of surface chemical potentials. From the foregoing equations it is evident that there are a number of possible definitions of the chemical potential in systems which contain a surface. Thus from (1.13)

$$\mu_i = \left(\frac{\partial A}{\partial n_i}\right)_{T,V,n_j,a}. \tag{1.27}$$

Correspondingly, a surface chemical potential may be defined from (1.12) as

$$\mu_i^\sigma = \left(\frac{\partial A^\sigma}{\partial n_i^\sigma}\right)_{T,V^\sigma,n_j^\sigma,a} \tag{1.28}$$

but if the surface is in equilibrium with the bulk

$$\mu_i = \mu_i^\sigma. \tag{1.29}$$

μ_i^σ may also be written in the equivalent form $(\partial A/\partial n_i^\sigma)_{T,V,n_i^\alpha,n_i^\beta,n_j^\sigma,a}$ or, as is evident from (1.27) and (1.29), $(\partial A/\partial n_i)_{T,V,n_j,a}$. This equivalence arises because the free energy change on addition of material is independent of the phase (α, β or interface) to which it is added. It is, of course, assumed that no concentration change results from the addition.

The chemical potential may also be expressed in terms of the Gibbs functions. Thus, from (1.20) and (1.19) respectively

$$\left.\begin{aligned} \mu_i &= \left(\frac{\partial \mathscr{G}}{\partial n_i}\right)_{T,p,n_j,a} \\ \text{and}\qquad \mu_i^\sigma &= \left(\frac{\partial \mathscr{G}^\sigma}{\partial n_i^\sigma}\right)_{T,p,n_j^\sigma,a}. \end{aligned}\right\} \tag{1.30}$$

Alternatively, from (1.23) and (1.22) respectively,

$$\left.\begin{aligned} \mu_i &= \left(\frac{\partial G}{\partial n_i}\right)_{T,p,n_j,\gamma} \\ \text{and}\qquad \mu_i^\sigma &= \left(\frac{\partial G^\sigma}{\partial n_i^\sigma}\right)_{T,p,n_j^\sigma,\gamma}. \end{aligned}\right\} \tag{1.31}$$

Consider the expression for μ_i^σ as obtained from (1.12),

$$\mu_i^\sigma = \left(\frac{\partial A^\sigma}{\partial n_i^\sigma}\right)_{T,V^\sigma,n_j^\sigma,a}.$$

It can be shown by the use of the properties of partial derivatives that this equation may be rewritten as

$$\mu_i^\sigma = \left(\frac{\partial A^\sigma}{\partial n_j^\sigma}\right)_{T, V^\sigma, n_j^\sigma, \gamma} - \left(\frac{\partial A^\sigma}{\partial \mathcal{A}}\right)_{T, V^\sigma, n_i^\sigma} \left(\frac{\partial \mathcal{A}}{\partial n_j^\sigma}\right)_{T, V^\sigma, n_j^\sigma, \gamma}. \tag{1.32}$$

But $(\partial A^\sigma/\partial \mathcal{A})_{T, V^\sigma, n_i^\sigma}$ is a definition of γ and $(\partial \mathcal{A}/\partial n_j^\sigma)_{T, V^\sigma, n_j^\sigma, \gamma}$ is the partial molar surface area, a_i, of component i. It is therefore possible to write (1.32) as

$$\mu_i^\sigma = \zeta_i - \gamma a_i, \tag{1.33}$$

where ζ_i is defined as

$$\zeta_i = \left(\frac{\partial A^\sigma}{\partial n_i^\sigma}\right)_{T, V^\sigma, n_j^\sigma, \gamma}. \tag{1.34}$$

The quantity ζ_i has sometimes been referred to as a surface chemical potential, but this terminology is not recommended since ζ_i is clearly not equal to μ_i or μ_i^σ. For a system at equilibrium the situation is summarised by

$$\mu_i^\sigma - \mu_i^l = 0$$

and

$$\zeta_i - \mu_i^l = \gamma a_i, \tag{1.35}$$

where μ_i^l represents the chemical potential of i in the bulk liquid. It will be convenient to use both μ_i^σ and ζ_i in later sections.

1.8. The Gibbs adsorption equation. When adsorption takes place at an interface, γ changes. One of the main objectives of surface chemistry is to determine the amount of material which is adsorbed at an interface. The Gibbs adsorption equation is a thermodynamic expression which relates the surface concentration (or surface excess) of a species to both γ and the bulk activity or fugacity of the adsorbate. Thus, in systems where γ is directly and simply measurable (i.e. liquid–liquid and liquid–vapour systems), the Gibbs adsorption equation may be used to determine the surface concentration. In other systems, where the surface concentration can be measured directly, but γ cannot (e.g. solid–gas systems), the Gibbs equation can be used to calculate the lowering of γ (i.e. the spreading pressure, π) which would not otherwise be available.

The adsorption equation can be derived using either of the models for the surface (§ 1.6) and the derivation takes the same form in either case. It is only necessary to recall that if the Gibbs approach is used $V^\sigma = 0$ and all extensive thermodynamic properties of the surface are excess rather than total quantities.

It is convenient to start with (1.14) which, it will be remembered, is valid for either surface convention and could equally well be written with superscript s. On differentiation this equation becomes

$$dA^\sigma = -p\,dV^\sigma - V^\sigma dp + \gamma\, d\mathscr{A} + \mathscr{A}\, d\gamma + \sum_i \mu_i dn_i^\sigma + \sum_i n_i^\sigma\, d\mu_i. \quad (1.36)$$

The comparison of (1.12) with (1.36) gives

$$S^\sigma dT - V^\sigma dp + \mathscr{A}\, d\gamma + \sum_i n_i^\sigma\, d\mu_i = 0. \quad (1.37)$$

For all practical purposes the term $V^\sigma dp$ is negligible or zero and so, for constant temperature,

$$\mathscr{A}\, d\gamma + \sum_i n_i^\sigma\, d\mu_i = 0. \quad (1.38)$$

Dividing (1.38) by $\mathscr{A}$,

$$d\gamma + \sum_i \frac{n_i^\sigma}{\mathscr{A}}\, d\mu_i = 0. \quad (1.39)$$

$n_i^\sigma/\mathscr{A}$ is written Γ_i and is either the surface concentration or, if the Gibbs approach is used, the surface excess of component i. Equation (1.39) may now be written

$$-d\gamma = \sum_i \Gamma_i\, d\mu_i \quad \blacktriangleleft\mathbf{(1.40)}$$

which is a general form of the Gibbs adsorption equation. It can equally well be derived starting from the differential expression for U^σ, $\mathscr{G}^\sigma$ and G^σ ((1.8), (1.19) and (1.22) respectively). The equation finds wide application in the study of adsorption, and some ways in which it can be used will now be outlined.

Consider adsorption from a two component liquid mixture at the liquid–vapour interface. The Gibbs adsorption equation (1.40) becomes

$$-d\gamma = \Gamma_1 d\mu_1 + \Gamma_2 d\mu_2, \quad (1.41)$$

where subscripts 1 and 2 refer to the two components of the mixture. The terms Γ_1 and Γ_2 in (1.41) are both unknown and various procedures are adopted in order to assign values to these quantities.

If the Gibbs approach is to be used, so that Γ_1 and Γ_2 are surface excesses, the usual convention is to place the dividing surface (SS' in fig. 1.3) such that $\Gamma_1 = 0$, and then the excess of 2 is written $\Gamma_2^{(1)}$ to denote this choice. The term $\Gamma_2^{(1)}$ may be called the *relative adsorption*. Thus,

$$-d\gamma = \Gamma_2^{(1)} d\mu_2. \quad (1.42)$$

Since $$\mu_2 = \mu_2^0 + RT \ln \boldsymbol{a}_2,$$

where μ_2^0 is the standard chemical potential of component 2 in the solution and $\boldsymbol{a}_2$ its activity, $d\mu_2$ is given by

$$d\mu_2 = RT\, d\ln \boldsymbol{a}_2$$

and (1.42) becomes

$$\Gamma_2^{(1)} = -\frac{1}{RT}\frac{d\gamma}{d\ln \boldsymbol{a}_2}. \qquad ◀(\mathbf{1.43})$$

If, on the other hand, the model is used in which the surface is assumed to have finite thickness, Γ_1 and Γ_2 in (1.41) are total surface concentrations (written Γ_1^s and Γ_2^s) and the following procedure is usually adopted. The Gibbs–Duhem relationship for a two component bulk solution at constant temperature and pressure is

$$x_1 d\mu_1 = -x_2 d\mu_2,$$

where x_1 and x_2 are the mole fractions of components 1 and 2. Equation (1.41) can now be written

$$-d\gamma = \left[\Gamma_2^s - \frac{x_2}{x_1}\Gamma_1^s\right] d\mu_2 \qquad (1.44)$$

so that

$$\Gamma_2^s - \frac{x_2}{x_1}\Gamma_1^s = -\frac{1}{RT}\frac{d\gamma}{d\ln \boldsymbol{a}_2}. \qquad (1.45)$$

It can readily be shown (Guggenheim, 1967*c*) that the magnitude of the left-hand side of (1.45) is independent of the positions of the planes *AA′* and *BB′* (fig. 1.2, §1.6) although this is not true for Γ_1^s and Γ_2^s individually.

The comparison of (1.45) with (1.43) shows that

$$\Gamma_2^{(1)} = \Gamma_2^s - \frac{x_2}{x_1}\Gamma_1^s. \qquad ◀(\mathbf{1.46})$$

The right-hand side of (1.46), like $\Gamma_2^{(1)}$, is a surface excess concentration. Thus, it is the number of moles of 2 per unit area of surface less the number of moles of 2 in that part of the bulk region which contains the same number (Γ_1^s) of moles of 1.

For dilute solutions of strongly adsorbing substances, many of which are of considerable practical interest, $x_2\Gamma_1^s/x_1$ is negligibly small and $\Gamma_2^{(1)} \simeq \Gamma_2^s$. In general, however, it is not possible to assign values to both Γ_1^s and Γ_2^s without making a non-thermodynamic assumption. Thus, if it is supposed that the surface phase is a monomolecular layer with respect to both components, then it is possible to write

$$\Gamma_1^s a_1 + \Gamma_2^s a_2 = 1, \qquad (1.47)$$

where a_1 and a_2 are the partial molar surface areas of components 1 and 2 and are assumed constant. If values may be assigned to a_1 and a_2 and if $(\Gamma_2^s - x_2\Gamma_1^s/x_1)$ has been determined experimentally, then both Γ_1^s and Γ_2^s can be calculated. However, it should be emphasised that as a rule Γ_1^s and Γ_2^s will not be equal to $\Gamma_1^{(2)}$ and $\Gamma_2^{(1)}$ respectively, as is obvious from (1.46).

It has already been indicated that the Gibbs adsorption equation may be applied to the study of solid–gas systems (see §5.5). The equation also appears, in a different guise, in the treatment of solid–liquid interfaces (§6.8 and §6.12) and is used for the interconversion of surface equations of state and adsorption isotherms (see §1.10 and §1.11).

1.9. Some general remarks on surface equations of state and adsorption isotherms. There are two basic pieces of information which are sought in a study of physical adsorption. The first is the way in which the extent of adsorption of a species depends on its bulk concentration or pressure, and the second is the state of the adsorbate at the surface. These two questions are closely interrelated since the amount of a species adsorbed depends on its state at the surface.

An *adsorption isotherm* is the mathematical expression which relates the bulk pressure or concentration of an adsorbing species to its surface concentration, at constant temperature. It may be written quite generally as

$$p = Kf(\Gamma), \tag{1.48}$$

where p is the pressure or concentration, K is a proportionality constant and Γ is a surface concentration. The expression which describes the behaviour of the molecules in the surface film is termed a *surface equation of state*, and relates the spreading pressure to the surface concentration. Such an equation is a two-dimensional analogue of the three-dimensional equation of state and may be written (cf. §1.3)

$$\pi = \gamma_0 - \gamma = RTf'(\Gamma), \tag{1.49}$$

where T is the absolute temperature and R the gas constant. The use of Γ in both (1.48) and (1.49) raises the question as to whether surface excess or total surface concentration is the more appropriate. As the surface tension is directly related, through the Gibbs adsorption equation, to the surface excess concentration, Γ^σ is the more appropriate choice. In the majority of systems where the surface pressure is considered, such as insoluble or strongly adsorbed monomolecular films, there is no appreciable difference between Γ^s and Γ^σ (see §1.10). The simplicity of the statistical treatment which follows is a direct consequence of this fact. However, it should be remembered throughout that, strictly speaking, π is determined by Γ^σ.

A surface equation of state is concerned only with *lateral* motions and interactions; an adsorption isotherm is concerned in addition with the interactions *normal* to the surface between adsorbate and adsorbent. It is possible to interconvert the two expressions by use of the Gibbs adsorption equation.

The general procedure in an investigation into the nature of an adsorbed film is to take an isotherm or equation of state pertaining to a specific model and to test the equation with experimental results. It is often possible to reproduce the shape of an experimental isotherm with a theoretical isotherm for a particular temperature. This, however, is not a sufficient test of a theory, as it is also necessary to be able to predict the effect of temperature on the adsorption.

The effect of heterogeneity of an adsorbent surface on the adsorption and surface behaviour of a species will be ignored in this chapter for the sake of simplicity. In addition the equations for multilayers are omitted at this stage since multilayer formation is mainly associated with adsorption on solid surfaces and will be discussed in chapter 5.

1.10. The two-dimensional perfect gas (i.e. an ideal non-localised monolayer). The equation of state for such a film is

$$\pi\mathcal{A} = NkT, \tag{1.50}$$

where $\mathcal{A}$ is the area of the film and N the number of molecules contained in it. This equation has often been assumed merely by analogy with the perfect gas law and may be obtained in a number of ways. The derivation which follows involves the use of statistical thermodynamics in a manner similar to that of Fowler & Guggenheim (1939*a*). It may reasonably be objected that such a powerful method is unnecessary to solve such a simple problem and, indeed, that it tends to obscure rather than clarify the issues. Nevertheless, the use of statistical thermodynamics becomes increasingly necessary in the more complex systems to be discussed later, and the derivation of the two-dimensional ideal gas equation constitutes an attractively simple introduction to the basic principles of the statistical approach.

The translational partition function, $l(T)$, for a vanishingly small molecule of mass m, confined to a surface of area $\mathcal{A}$, at a temperature T, is

$$l(T) = \frac{2\pi mkT}{h^2}\mathcal{A}, \tag{1.51}$$

where h is Planck's constant and π is the mathematical constant. The partition function for the internal degrees of freedom of an adsorbed molecule, including vibrations normal to the surface, may be written $j^{\mathrm{ads}}(T)$. When referred to an energy zero corresponding to the ground internal level of an isolated molecule in the gas phase this partition function becomes $j^{\mathrm{ads}}(T)\exp(V/kT)$. Here V is the minimum energy needed to remove the molecule, in its lowest energy state, from the surface to the bulk. The complete partition function for the molecule at the surface is

$$\mathfrak{z}(T) = \frac{2\pi mkT}{h^2}\mathcal{A}j^{\mathrm{ads}}(T)\exp\left(\frac{V}{kT}\right). \tag{1.52}$$

That part of the Helmholtz free energy of the surface which arises as a consequence of the adsorbed molecules is denoted A^{ads}. The total Helmholtz free energy of the surface is made up of A^{ads} together with contributions from the underlying molecules. For the present purposes only A^{ads} is required.

If the surface monolayer consists of N indistinguishable mutually non-interacting molecules, the partition function for the monolayer, Z^{ads}, is given by

$$Z^{\mathrm{ads}} = \frac{1}{N!}\{\mathfrak{z}(T)\}^N. \tag{1.53}$$

The expressions for A^{ads} and the surface pressure of the adsorbed molecules in terms of partition functions are analogous to those for bulk phases, i.e.

$$A^{\mathrm{ads}} = -kT\ln Z^{\mathrm{ads}} \qquad \blacktriangleleft\mathbf{(1.54)}$$

and

$$\pi = -\left(\frac{\partial A^{\mathrm{ads}}}{\partial \mathcal{A}}\right)_{T,V,N}$$

$$= kT(\partial \ln Z^{\mathrm{ads}}/\partial\mathcal{A})_{T,V,N}. \qquad \blacktriangleleft\mathbf{(1.55)}$$

The combination of (1.53) and (1.54), using Stirling's approximation, yields

$$A^{\mathrm{ads}} = -NkT\ln\mathfrak{z}(T) + NkT(\ln N - 1)$$

and from (1.51) and (1.52) this becomes

$$A^{\mathrm{ads}} = NkT\ln\left\{\frac{N}{\mathcal{A}}\frac{h^2}{2\pi mkT}\frac{1}{j^{\mathrm{ads}}(T)\exp(V/kT)}\right\} - NkT. \tag{1.56}$$

On differentiation with respect to $\mathcal{A}$, (1.56) gives, in combination with (1.55) the required equation of state

$$\pi\mathcal{A} = NkT. \qquad \blacktriangleleft\textbf{(1.57)}$$

The adsorption isotherm may also be derived from A^{ads} in the following way. The chemical potential μ^{ads} of the adsorbate is given by

$$\mu^{\text{ads}} = \left(\frac{\partial A^{\text{ads}}}{\partial N}\right)_{T,V,a} = \frac{\partial}{\partial N}\{-NkT\ln\mathfrak{z}(T)+NkT(\ln N-1)\}$$
$$= -kT\ln\mathfrak{z}(T)+kT\ln N. \qquad (1.58)$$

The bulk chemical potential μ^g for a perfect gas is (Fowler & Guggenheim, 1939*b*)

$$\mu^g = kT\ln\left\{\frac{p}{kT}\frac{h^3}{(2\pi mkT)^{\frac{3}{2}}j^g(T)}\right\}, \qquad (1.59)$$

where p is the pressure of the adsorbate and $j^g(T)$ is the internal partition function for an adsorbate molecule in the gas phase.

The chemical potentials of the adsorbate in the surface and bulk phases are equal at adsorption equilibrium and the comparison of (1.58) and (1.59) gives the adsorption isotherm

$$p = \left\{\frac{kT}{\mathcal{A}}\frac{(2\pi mkT)^{\frac{1}{2}}}{h}\frac{j^g(T)}{j^{\text{ads}}(T)}\frac{1}{\exp(V/kT)}\right\}N,$$

i.e.

$$p = KN, \qquad \blacktriangleleft\textbf{(1.60)}$$

where

$$K = \frac{kT}{\mathcal{A}}\frac{(2\pi mkT)^{\frac{1}{2}}}{h}\frac{j^g(T)}{j^{\text{ads}}(T)}\frac{1}{\exp(V/kT)}.$$

Equation (1.60) is analogous to Henry's law and shows that for the adsorption from a perfect gas to a perfect non-localised monolayer, the amount adsorbed is directly proportional to the pressure. The constant K, which contains the potential energy term V describes adsorbate–adsorbent interactions normal to the surface and it will be shown in §1.14 that K may be expressed in terms of a standard free energy change accompanying adsorption.

As indicated earlier, it is possible to convert an equation of state to the corresponding adsorption isotherm with the aid of the Gibbs adsorption equation. The latter is written, for the adsorption of a single component, as

$$-\mathrm{d}\gamma = \Gamma^\sigma\,\mathrm{d}\mu. \qquad (1.61)$$

Γ^σ is referred to a Gibbs dividing surface located such that the surface excess of the other phase (e.g. solid) is zero; i.e. at the solid surface. Suppose that the 'surface phase' convention is used (fig. 1.2). The plane AA' is located at the solid surface, and the plane BB' is placed in the gas phase close to the surface. Then for low pressures of gas Γ^s is to a good approximation equal to Γ^σ, and (1.61) may be written $-\mathrm{d}\gamma = \Gamma^s \mathrm{d}\mu$. Since

$$\pi = \gamma_0 - \gamma, \quad \mathrm{d}\pi = -\mathrm{d}\gamma$$

and

$$\mathrm{d}\pi = \Gamma^s \mathrm{d}\mu \tag{1.62}$$

$$= \Gamma^s RT\,\mathrm{d}\ln p. \tag{1.63}$$

Since

$$\Gamma^s = N/\mathcal{A}N_A, \tag{1.64}$$

where N_A is the Avogadro number, (1.63) becomes

$$\mathrm{d}\pi = \frac{NkT}{\mathcal{A}}\,\mathrm{d}\ln p. \tag{1.65}$$

On differentiation (1.57) becomes

$$\mathrm{d}\pi = \frac{kT}{\mathcal{A}}\,\mathrm{d}N. \tag{1.66}$$

By comparison of (1.65) and (1.66), and integration

$$p = KN. \tag{1.67}$$

This equation is identical to (1.60) and the constants K are evidently the same.

The advantage of deriving the isotherm statistically is that K is given explicitly in terms of molecular parameters, whereas in the conversion of the equation of state K appears only as an integration constant.

1.11. Non-ideal non-localised monolayers. The above treatment of a perfect two-dimensional gas assumes effectively point molecules which have no lateral attractions and is useful only for very dilute films. For more concentrated monolayers, however, the size of a molecule may not be negligible compared to the area ($\mathcal{A}/N$) of surface available to it. The equation of state which, when lateral attractions are absent, often accounts well for the behaviour of such a system is that proposed by Volmer (1925),

$$\pi(\mathcal{A} - N\dot{a}_0) = NkT, \tag{◄1.68}$$

where $\dot{a}_0$ is known as the co-area of a molecule. Just as the equation of state (1.58) is consistent with the translational partition function (1.51), so (1.68) is consistent with

$$l(T) = \frac{2\pi mkT}{h^2}(\mathcal{A} - N\dot{a}_0). \qquad (1.69)\dagger$$

The complete partition function for a molecule is then

$$\mathfrak{z}(T) = \frac{2\pi mkT}{h^2}(\mathcal{A} - N\dot{a}_0)j^{\text{ads}}(T)\exp\left(\frac{V}{kT}\right) \qquad (1.70)$$

from which (1.68) may be derived by the procedure given in §1.10.

The adsorption isotherm which corresponds to (1.68) may also be derived starting from (1.70); alternatively, it may be obtained from (1.68) by the use of the Gibbs adsorption equation. Both procedures are relatively simple but, for the sake of brevity, only the latter derivation will be given (de Boer, 1968).

If (1.68) is divided by N and differentiated with respect to $\dot{a}(=\mathcal{A}/N)$, the area per molecule, the following equation is obtained

$$d\pi = -\frac{kT}{(\dot{a}-\dot{a}_0)^2}d\dot{a}. \qquad (1.71)$$

The combination of (1.63) and (1.64) gives

$$d\pi = \frac{kT}{\dot{a}}d\ln p \qquad (1.72)$$

and therefore

$$d\ln p = -\frac{\dot{a}\,d\dot{a}}{(\dot{a}-\dot{a}_0)^2}. \qquad (1.73)$$

On integration, (1.73) yields

$$\ln p = -\left\{\ln(\dot{a}-\dot{a}_0) - \frac{\dot{a}_0}{\dot{a}-\dot{a}_0}\right\} + \text{constant}, \qquad (1.74)$$

† The following remarks on the physical significance of $\dot{a}_0$ may be helpful. In a two-dimensional gas the molecules cannot approach more closely to each other than the sum (r^*) of their equivalent hard sphere radii will permit, and around the centre of each molecule there is an area, πr^{*2}, from which the centres of all other molecules are excluded. This excluded area is common to two molecules, and therefore for the whole gas the excluded area per molecule is only half this, i.e. $\pi r^{*2}/2$. The so-called free area for the gas molecules is therefore $(\mathcal{A} - N\dot{a}_0)$ where $\dot{a}_0 = \pi r^{*2}/2$ and this free area is introduced into the partition function in place of the actual area $\mathcal{A}$ used for a perfect gas. It should be emphasised that this interpretation of $\dot{a}_0$ is valid only when the gas is dilute.

which can be written

$$p = K_1' \frac{1}{\dot{a}-\dot{a}_0} \exp \frac{\dot{a}_0}{\dot{a}-\dot{a}_0}$$

i.e.

$$p = K_1 \frac{\dot{a}_0}{\dot{a}-\dot{a}_0} \exp \frac{\dot{a}_0}{\dot{a}-\dot{a}_0}, \qquad ◀(\mathbf{1.75})$$

where $K_1 = K_1'/\dot{a}_0$. The quantities $\dot{a}$ and $\dot{a}_0$ in (1.75) are sometimes expressed in terms of *surface coverage*, θ, where $\theta = \dot{a}_0/\dot{a}$. In terms of θ therefore, (1.75) takes the form

$$p = K_1 \frac{\theta}{1-\theta} \exp \frac{\theta}{1-\theta}. \qquad (1.76)$$

When lateral attractions within the monolayer become significant, the equation of state has often been assumed to take the form of the two-dimensional van der Waals equation

$$\left(\pi + \frac{N^2\alpha}{\mathcal{A}^2}\right)(\mathcal{A} - N\dot{a}_0) = NkT, \qquad ◀(\mathbf{1.77})$$

where α is a two-dimensional van der Waals constant which allows for lateral attractions in the film. The complete partition function for a molecule which is consistent with (1.77) is

$$\mathfrak{z}(T) = \frac{2\pi mkT}{h^2}(\mathcal{A} - N\dot{a}_0)\, j^{\mathrm{ads}}(T) \exp\left(\frac{V}{kT}\right) \exp\left(\frac{\alpha N}{\mathcal{A}kT}\right). \qquad (1.78)$$

The adsorption isotherm, which can be obtained from (1.78), is (Hill, 1952)

$$p = K_2 \frac{\theta}{1-\theta} \exp\left(\frac{\theta}{1-\theta} - \frac{2\alpha\theta}{\dot{a}_0 kT}\right). \qquad ◀(\mathbf{1.79})$$

Just as for K and K_1 in (1.60) and (1.76) respectively, K_2 is given by the statistical derivation explicitly in terms of molecular parameters. Expression (1.79) is derived on the assumption that the bulk gas behaves ideally; (1.77) and (1.79) are useful since they describe, if only qualitatively, two-dimensional condensation in monolayers. This phenomenon is observed experimentally for monolayers both on liquid and solid surfaces (see §§3.7 and 5.10).

1.12. Ideal localised monolayers. The adsorption equation derived on the assumption of an ideal localised monolayer has the form of the well-known Langmuir isotherm. This was originally obtained in 1918 by the use of a kinetic approach. The equation was used widely to describe experimental adsorption isotherms for gases and vapours on solids but it has since been realized that the model assumed is inapplicable to many of these systems. The model is apparently more appropriate to systems which exhibit chemisorption. Perhaps more important for present purposes, the model has been used as the starting point for the derivation of isotherms for multilayer adsorption of gases on solids (chapter 5). A statistical derivation of the Langmuir adsorption isotherm is given below; this is a slight modification of that given by Fowler & Guggenheim (1939*c*).

The system to be discussed is that of a homogeneous adsorbent surface in contact with a single component gas. The surface is supposed to be made up of N_s equivalent adsorption sites, N of which are assumed to be occupied by mutually non-interacting adsorbed gas molecules; each site can accommodate only one adsorbed molecule. $N_s - N$ then represents the number of vacant sites. For a molecule at the surface, a partition function $\mathfrak{z}(T)$ for internal degrees of freedom, including vibrations about the mean position in a site, may be written (see §1.10)

$$\mathfrak{z}(T) = j^{\text{ads}}(T) \exp\left(\frac{V}{kT}\right). \tag{1.80}$$

The complete partition function of the monolayer, Z^{ads}, is then

$$Z^{\text{ads}} = \begin{Bmatrix} \text{number of distinguishable ways of arranging } N \\ \text{molecules over } N_s \text{ sites} \end{Bmatrix} [\mathfrak{z}(T)]^N, \tag{1.81}$$

where the configurational term is

$$\frac{N_s!}{N!(N_s - N)!}. \tag{1.82}$$

The Helmholtz free energy of the monolayer, A^{ads}, is given by

$$A^{\text{ads}} = -kT \ln Z^{\text{ads}}, \tag{1.83}$$

and therefore

$$A^{\text{ads}} = kT[-N_s \ln N_s + N \ln N + (N_s - N) \ln (N_s - N) - N \ln \mathfrak{z}(T)]. \tag{1.84}$$

The chemical potential, μ^{ads}, of the adsorbate in the monolayer is now given by

$$\begin{aligned} \mu^{\text{ads}} = \left(\frac{\partial A^{\text{ads}}}{\partial N}\right)_{T,V,a} &= kT \ln \frac{N}{N_s - N} - kT \ln \mathfrak{z}(T) \\ &= kT \ln \frac{\theta}{1-\theta} - kT \ln \mathfrak{z}(T), \end{aligned} \tag{1.85}$$

where $\theta = N/N_s$. The chemical potential of the adsorbate in the gas phase, assuming ideality, is given by

$$\mu^g = kT\ln\left\{\frac{p}{kT}\frac{h^3}{(2\pi mkT)^{\frac{3}{2}}j^g(T)}\right\}, \tag{1.86}$$

where $j^g(T)$ is the internal partition function for an adsorbate molecule in the gas phase. A comparison of (1.85) and (1.86) shows that at adsorption equilibrium

$$p = K_3\frac{\theta}{1-\theta}, \tag{◀1.87}$$

where
$$K_3 = \frac{kT}{j^{\text{ads}}(T)\exp(V/kT)}\frac{(2\pi mkT)^{\frac{3}{2}}j^g(T)}{h^3}. \tag{1.88}$$

Equation (1.87) is a form of the Langmuir adsorption isotherm.

In (1.87) $\theta/(1-\theta)$ is a configurational or entropic term which arises from the assumption of a localised monolayer. K_3, like K, K_1 and K_2 in the isotherms already discussed, is related to the difference in the standard chemical potentials of the adsorbate in the bulk gas and the surface. This can be made clear in the following way.

For one mole of adsorbate, (1.85) may be written as

$$\mu^{\text{ads}} = RT\ln\frac{\theta}{1-\theta} - RT\ln \mathfrak{z}(T). \tag{1.89}$$

It should be noted that the same symbol has been used for both molar and molecular chemical potentials although the two quantities are obviously not the same. In (1.89) the term in $\theta/(1-\theta)$ is a configurational one and is, of course, a function of the amount of adsorbate on the surface. The term $RT\ln \mathfrak{z}(T)$, however, is independent of θ but dependent on temperature, and may therefore be written as a standard chemical potential $-\mu^{0,\text{ads}}$. Thus, (1.89) becomes

$$\mu^{\text{ads}} = \mu^{0,\text{ads}} + RT\ln\frac{\theta}{1-\theta}. \tag{1.90}$$

The chemical potential of the adsorbate in the (ideal) gas phase is written

$$\mu^g = \mu^{0,g} + RT\ln p, \tag{1.91}$$

where $\mu^{0,g}$ is the standard chemical potential in the gas phase.

From (1.90) and (1.91), at adsorption equilibrium

$$\mu^{0,\text{ads}} - \mu^{0,g} = RT\ln\frac{p}{\theta/(1-\theta)} \tag{1.92}$$

or
$$\exp\left(\frac{\mu^{0,\text{ads}}-\mu^{0,g}}{RT}\right) = \frac{p}{\theta/(1-\theta)}. \tag{1.93}$$

But from (1.87)
$$\frac{p}{\theta/(1-\theta)} = K_3. \tag{1.94}$$

It follows therefore that
$$K_3 = \exp\left(\frac{\mu^{0,\text{ads}}-\mu^{0,g}}{RT}\right). \quad \blacktriangleleft\text{(1.95)}$$

$\mu^{0,\text{ads}}-\mu^{0,g}$ may be written $\Delta_a\mu^0$ and is the change in standard chemical potential on adsorption. The standard states are unit pressure in the gas phase and half coverage on the surface. This latter state arises since when $\theta = \frac{1}{2}$, $\ln \theta/(1-\theta) = 0$, and therefore from (1.90) $\mu^{\text{ads}} = \mu^{0,\text{ads}}$. The relationship between the other *K*s and $\Delta_a\mu^0$ may be obtained in an analogous manner. This particular example has been chosen as the result will be needed in chapter 2. A further account of the relationship between the various *K*s occurring in adsorption isotherms, and $\Delta_a\mu^0$, is given in §1.14.

The surface equation of state for the ideal localised monolayer may be derived from (1.84) and is
$$\pi = \frac{kT}{\dot{a}_0}\ln\frac{\dot{a}}{\dot{a}-\dot{a}_0}. \quad \blacktriangleleft\text{(1.96)}$$

It is of interest to compare the forms of (1.87) and (1.96) with the corresponding expressions (1.76) and (1.68) for non-localised monolayers. The exponential term in (1.76) evidently arises from the non-localisation of the molecules in the monolayer.

When $\dot{a} \gg \dot{a}_0$, (1.96) obviously reduces to the form of the ideal two-dimensional gas equation (1.50); i.e. both localised and non-localised models give the same result.

1.13. Non-ideal localised monolayers. When lateral interactions between adsorbate molecules exist in a localised film the equation of state and adsorption isotherm may be written, respectively, as (Ross & Olivier, 1964*a*; Fowler & Guggenheim, 1939*d*)
$$\pi = \frac{kT}{\dot{a}_0}\ln\frac{\dot{a}}{\dot{a}-\dot{a}_0}+\frac{cV\dot{a}_0}{2\dot{a}^2} \tag{1.97}$$

and
$$p = K_4\frac{\theta}{1-\theta}\exp\frac{cV\theta}{kT}. \tag{1.98}$$

In these equations *V* is the energy of interaction of a pair of molecules on nearest neighbour sites on the surface and c is the number of nearest neighbour sites per site. The partition function for a molecule on the surface which leads to (1.97) and (1.98) is

$$\mathfrak{z}(T) = j^{\text{ads}}(T) \exp (V/kT) \exp (-cV\dot{a}_0 N/2\mathcal{A}kT). \qquad (1.99)$$

In (1.97)–(1.99) an attraction between molecules corresponds to a negative value of V. This is the more usual convention, and may be contrasted with that commonly used in (1.77)–(1.79) for non-localised monolayers, where attraction corresponds to a positive value of α.

The model assumed is only approximate in that the molecules are supposed to be randomly distributed over the adsorption sites even though interactions are present. This point is further discussed in the section dealing specifically with monolayers in gas–solid systems (chapter 5).

Like (1.77) and (1.79), (1.97) and (1.98) predict that two-dimensional condensation should occur at sufficiently low temperatures.

1.14. Standard chemical potential of adsorption from adsorption isotherms. The change in chemical potential, $\Delta_a\mu$, on the adsorption of one mole of a species, is a function of the extent of adsorption, and the molar entropy of the adsorbate changes with coverage and with bulk pressure or concentration. In addition, if lateral interactions are present in the adsorbed film, these will also change with coverage and $\Delta_a\mu$ will again be affected. $\Delta_a\mu$ is therefore composed of contributions from interactions normal to and parallel to the surface.

Suppose that one mole of a gas at temperature T and originally at its standard pressure, $p^{\ominus}$, is adsorbed isothermally. If the adsorbed state is defined by the equilibrium pressure, p, and surface coverage, θ, $\Delta_a\mu$ is (Ross & Olivier, 1964*b*)

$$\Delta_a\mu = RT \ln (p/p^{\ominus}). \qquad (1.100)$$

It is possible to substitute for p in (1.100) from the isotherms of §1.10 to §1.13. In general the isotherm is $p = Kf(\theta)$ (cf. (1.48)) so that (1.100) becomes

$$\Delta_a\mu = RT \ln \frac{K}{p^{\ominus}} + RT \ln f(\theta). \qquad (1.101)$$

For example, in the case of an interacting non-localised monolayer, where the adsorption isotherm is given by (1.79), (1.101) now becomes

$$\Delta_a\mu = RT \ln \frac{K_2}{p^{\ominus}} + RT \ln \frac{\theta}{1-\theta} + RT\left(\frac{\theta}{1-\theta} - \frac{2\alpha\theta}{\dot{a}_0 kT}\right). \qquad (1.102)$$

If a comparison of $\Delta_a \mu$ in different systems is to be made then clearly this quantity must be referred to a standard value of θ and of bulk pressure (or concentration) of adsorbate. The quantity of interest then becomes a *standard* chemical potential change, $\Delta_a \mu^0$.

One convenient standard state for the surface is half coverage, with lateral interactions absent; for the bulk phase, unit pressure (assumed ideal) can be used. When such standard states are employed (1.102) takes the form

$$\Delta_a \mu^0 = RT \ln K_2 + RT$$
$$= RT(\ln K_2 + 1).$$

Thus
$$K_2 = \exp\left(\frac{\Delta_a \mu^0}{RT} - 1\right) \qquad \blacktriangleleft\textbf{(1.103)}$$

and (1.79) can be written

$$\frac{\theta}{1-\theta} \exp\left(\frac{\theta}{1-\theta} - \frac{2\alpha\theta}{\dot{a}_0 kT}\right) = p \exp\left(1 - \frac{\Delta_a \mu^0}{RT}\right). \qquad (1.104)$$

K_3, which appears in the Langmuir isotherm (§1.12) has been expressed in terms of the standard chemical potential of adsorption in (1.95). The same result is obtained by the above procedure. When the standard states are taken as $\theta = \frac{1}{2}$ and $p^{\ominus} = 1$, the combination of (1.87) and (1.100) yields

$$\Delta_a \mu^0 = RT \ln K_3$$

or
$$K_3 = \exp\left(\frac{\Delta_a \mu^0}{RT}\right) \qquad \blacktriangleleft\textbf{(1.105)}$$

which is equivalent to (1.95).

1.15. Testing isotherms and equations of state. A common procedure is to convert the isotherm or equation of state to a linear form and plot experimental data accordingly. It is possible to obtain surface pressures for liquid–vapour and liquid–liquid systems directly from surface or interfacial tension data. In addition, from the same data, the surface concentration of adsorbate can be calculated by use of the Gibbs adsorption equation (chapter 3). It is a simple matter, therefore, to test a surface equation state in such systems. It is equally simple to test the applicability of an isotherm. It is not, however, possible to obtain a direct measure of the surface pressure at solid surfaces, and so here it is usual to test the isotherm only.

References

Aston, J. G. & Fritz, J. J. (1959). *Thermodynamics and Statistical Thermodynamics* (Wiley): pp. 241 *et seq.*

Butler, J. A. V. (1951). *Chemical Thermodynamics*, 4th edition (Macmillan): chapter 21.

de Boer, J. H. (1968). *The Dynamical Character of Adsorption*, 2nd edition (Clarendon Press): pp. 125–6.

Fowler, R. H. & Guggenheim, E. A. (1939). *Statistical Thermodynamics* (Cambridge University Press): (*a*) p. 423–4; (*b*) p. 427; (*c*) p. 426; (*d*) p. 430.

Gibbs, J. W. (1928). *Collected Works*, 2nd edition (Longmans, Green, New York): vol. 1, p. 219.

Guggenheim, E. A. (1967). *Thermodynamics*, 5th edition (North Holland Publishing Co.): (*a*) p. 45; (*b*) p. 47; (*c*) p. 208–9.

Hayward, D. O. & Trapnell, B. M. W. (1964). *Chemisorption*, 2nd edition (Butterworths).

Hildebrand, J. M. & Scott, R. C. (1950). *Solubility of Non-Electrolytes*, 3rd edition (Reinhold): chapter 21.

Hill, T. L. (1946). *J. Chem. Physics*, **14**, 441.

Hill, T. L. (1952). *Adv. Catalysis*, **4**, 211.

Kirkwood, J. G. & Oppenheim, I. (1961). *Chemical Thermodynamics* (McGraw-Hill): chapter 10.

Lewis, G. N. & Randall, M. (1961). *Thermodynamics*, 2nd edition, revised by K. S. Pitzer & L. Brewer (McGraw-Hill): chapter 29.

Ross, S. & Olivier, J. P. (1964). *On Physical Adsorption* (Interscience): (*a*) p. 17; (*b*) p. 19.

Volmer, M. (1925). *Z. phys. Chem.* **115**, 253.

2 Electrical potentials at interfaces

2.1. Introduction. At almost all interfaces there is a segregation of positive and negative charge in a direction normal to the phase boundary. The charges may be 'free', as are ions and electrons, or they may be associated in the form of dipolar molecules or polarised atoms. This normal segregation of charge may occur through the preferential adsorption of either positive or negative ions at the interface, through the adsorption and orientation of dipolar molecules or through a transfer of charge from one phase to another. Whatever the origin of the charge segregation, electrical potential differences are set up across the interface and an *electrical double layer* is formed.

The thermodynamic relationships for interfaces, such as those given in chapter 1, take account formally of the possible existence of interfacial potentials, but it must be remembered that for charged particles the chemical potentials which appear in these relationships are actually electrochemical potentials and that the system as a whole must remain electrically neutral.† Thermodynamic relationships, however, do not help in the understanding of the molecular structure of interfaces, and progress in this direction can only be achieved by an examination of the nature of electrochemical potentials. In the course of such an examination it is necessary to define and consider the measurement of certain electrical potentials.

In this chapter general aspects of interfacial potentials are discussed. In chapters 3 and 4 the results will be used in connection with specific types of system. More comprehensive treatments of some of the material in this chapter have been given by Parsons (1954), and by Guggenheim (1967). In method of presentation Parsons' approach has been adopted and for more detailed study his article particularly is recommended.

2.2. The definition and significance of interfacial potentials. Consider a system of two phases, α and β, which both contain ions of species i. Unless the system is at equilibrium the electrochemical potential $\tilde{\mu}_i$ of

† This point is well illustrated by the application of the Gibbs equation to electrode surfaces as described in chapter 4.

the ions will be different in the two phases and the work done in transferring an ion from the bulk of the α-phase to the bulk of the β-phase will be $(\tilde{\mu}_i^\beta - \tilde{\mu}_i^\alpha)$. It is convenient, although not necessarily meaningful, to regard this work as made up of two parts. The first of these arises from the difference in the nature of the material surrounding the ion in the two phases, i.e. from the change in the interactions of the ion with its neighbouring atoms and molecules when it is transferred. These interactions depend not only on the surroundings but also, obviously, on the ion itself. This part of the work of transfer is denoted $(\mu_i^\beta - \mu_i^\alpha)$. The second part of the work arises from the electrical potential difference $(\varphi^\beta - \varphi^\alpha)$ between the two bulk phases, and is given by $z_i e^-(\varphi^\beta - \varphi^\alpha)$. This part depends only on the charge $z_i e^-$ on the ion and not on the chemical nature of the ion or its material surroundings.

The total work of transfer of the ion can therefore be written

$$\tilde{\mu}_i^\beta - \tilde{\mu}_i^\alpha = (\mu_i^\beta - \mu_i^\alpha) + z_i e^-(\varphi^\beta - \varphi^\alpha)$$

or, more briefly,

$$\Delta^{\alpha\beta}\tilde{\mu}_i = \Delta^{\alpha\beta}\mu_i + z_i e^- \Delta^{\alpha\beta}\varphi. \quad \blacktriangleleft\mathbf{(2.1)}$$

If the transferred particle carries no net charge ($z_i = 0$) the work of transfer becomes simply $(\mu_i^\beta - \mu_i^\alpha)$, which is the chemical potential difference.

φ^α and φ^β are the electrical potentials in the interiors of the α- and β-phases. They are known as the *inner* potentials of the phases and their difference, $(\varphi^\beta - \varphi^\alpha)$, as the *Galvani* potential difference. $(\tilde{\mu}_i^\beta - \tilde{\mu}_i^\alpha)$ can be measured by the techniques of electrochemistry. However, the term $(\mu_i^\beta - \mu_i^\alpha)$ cannot be measured except in the trivial case of z_i equal to zero and therefore, in general, $(\varphi^\beta - \varphi^\alpha)$ is indeterminate. It has been pointed out by Gibbs (1928) and by Guggenheim (1929) that the splitting of electrochemical potential differences into 'chemical' and 'electrical' terms does not usually have any physical significance. In one instance only can $(\varphi^\beta - \varphi^\alpha)$ be measured. This is when the phases α and β have the same chemical composition. Then, $\mu_i^\beta = \mu_i^\alpha$ and therefore

$$\tilde{\mu}_i^\beta - \tilde{\mu}_i^\alpha = z_i e^-(\varphi^\beta - \varphi^\alpha). \quad \blacktriangleleft\mathbf{(2.2)}$$

This special case, however, is of very considerable practical value.

Further progress in the analysis of interfacial potentials depends very largely on the nature of the system. The term $(\varphi^\beta - \varphi^\alpha)$ is essentially the change in electrical potential experienced by a probe as it crosses the interfacial region between α and β. If, for example, either the α- or β-phase is a vacuum, φ for the remaining phase can be usefully broken down into two parts, one, ψ, which is due to the presence of an electrostatic charge on the surface of the phase and the other, χ, which is due to the presence of a dipolar charge distribution of any kind in the surface of the phase. Thus for the α-phase in a vacuum,

$$\varphi^\alpha = \psi^\alpha + \chi^\alpha. \qquad ◀\mathbf{(2.3)}$$

ψ^α is known as the *outer* potential of the phase. The reason for this name can be appreciated from the following consideration. Although, strictly speaking, the potential near a charged body in a vacuum varies continuously with the distance from its surface, it can be calculated from elementary electrostatic theory that, e.g. for a sphere 1 mm in radius, there is a region between approximately 10^{-4} and 10^{-2} mm from the surface in which the potential is effectively constant and equal to the average potential of the surface itself. At points closer to the surface than 10^{-4} mm the potential is likely to fall sharply owing to the imaging of the probe charge, and beyond 10^{-2} mm the potential should fall according to the long range coulombic interaction law. For larger spheres the outer limit increases. The effectively constant potential in the region between 10^{-4} and 10^{-2} mm outside the surface of the phase is denoted ψ^α. As the region in question and a point at a semi-infinite distance from the surface are in the same phase ψ^α may, according to (2.2), be measured by placing in the appropriate place electrodes of similar material.

χ^α is obviously the potential difference between the region where the potential is ψ^α and the interior of the α-phase; it is called the *surface* or *chi* potential. χ is considered to be positive when the potential increases from the outside to the inside of the phase. As φ^α is not measurable, it follows from (2.3) that χ^α also is not measurable.

The electrochemical potential for component *i* in phase α,

$$\tilde{\mu}_i^\alpha = \mu_i^\alpha + z_i e^- \varphi^\alpha \qquad (2.4)$$

may, if α is in a vacuum, be combined with (2.3) to give

$$\tilde{\mu}_i^\alpha = \mu_i^\alpha + z_i e^- (\psi^\alpha + \chi^\alpha). \qquad (2.5)$$

As $\tilde{\mu}_i^\alpha$ and ψ^α are well-defined quantities it is useful to rearrange (2.5) so as to obtain another well-defined potential; thus

$$\tilde{\mu}_i^\alpha - z_i e^- \psi^\alpha = \alpha_i^\alpha = \mu_i^\alpha + z_i e^- \chi^\alpha, \qquad ◀\mathbf{(2.6)}$$

where α_i^α is the *real* potential of *i* in the α-phase.

When α and β are both material phases, the Galvani potential difference $(\varphi^\beta - \varphi^\alpha)$ may be written

$$\varphi^\beta - \varphi^\alpha = \psi^\beta - \psi^\alpha + \chi^\beta - \chi^\alpha, \qquad \blacktriangleleft\textbf{(2.7)}$$

where the meaning of the terms is the same as that in (2.3). The outer potential difference $(\psi^\beta - \psi^\alpha)$ is often called the *Volta* potential difference and, like ψ^α, may be measured. The remaining terms in (2.7) are not measurable. It should be noted particularly that the terms on the right-hand side of (2.7) refer to the surfaces of α and β separately in a vacuum, and not to the $\alpha\beta$ interface. Indeed it is usually difficult, when α and β are material phases, to interpret the Galvani potential difference in terms of well-defined potentials at the $\alpha\beta$ interface. For instance, the variation of potential φ with distance from the interface is likely to be much more rapid in a material phase than in a vacuum (electrolyte solutions are a case in point, as will be seen below) and, in consequence, it is not so easy to define a potential analogous to an outer potential, as was done when one phase was in a vacuum. Nevertheless Parsons has pointed out that a definition might be achieved for a system consisting of a metal in a non-conducting dielectric where, apart from the influence of the dielectric constant, the potential would vary as in a vacuum. With regard to the surface potential, a term $\chi^{\alpha\beta}$ analogous to $(\chi^\beta - \chi^\alpha)$ can still be imagined for an interface between two material phases but, owing to molecular interaction across the interface, this term would almost certainly not be equal to $(\chi^\beta - \chi^\alpha)$.

From the way that it has been introduced it is obviously the surface potential χ that is primarily related to the molecular structure of a surface. Although this potential cannot be measured, the Volta potential, i.e. the difference between the outer potentials of two phases, can be measured and it transpires that one can often measure *changes* in χ when, for instance, thin films are spread or adsorbed at clean interfaces. For the purposes of illustration consider two metals, α and β, in contact with each other and under conditions where no electric current is flowing across their mutual interface. The electrochemical potentials of the electrons in each metal are equal, i.e.

$$\tilde{\mu}_e^\alpha = \tilde{\mu}_e^\beta \qquad (2.8)$$

and therefore, from (2.4) and (2.3)

$$\mu_e^\alpha - e^-\psi^\alpha - e^-\chi^\alpha = \mu_e^\beta - e^-\psi^\beta - e^-\chi^\beta, \qquad (2.9)$$

where the negative signs arise from the negative charge of the electrons. On rearrangement (2.9) becomes

$$\psi^\beta - \psi^\alpha = \Delta^{\alpha\beta}\psi = (\mu_e^\beta - \mu_e^\alpha)/e^- - (\chi^\beta - \chi^\alpha). \tag{2.10}$$

Now suppose that the Volta potential difference is measured first for the clean metal surfaces ($\Delta^{\alpha\beta}\psi_1$) and then in the presence of an adsorbing gas ($\Delta^{\alpha\beta}\psi_2$). Then

$$\Delta^{\alpha\beta}\psi_2 - \Delta^{\alpha\beta}\psi_1 = [(\mu_e^\beta)_2 - (\mu_e^\beta)_1]/e^- - [(\mu_e^\alpha)_2 - (\mu_e^\alpha)_1]/e^- \\ -(\chi_2^\beta - \chi_1^\beta) + (\chi_2^\alpha - \chi_1^\alpha). \tag{2.11}$$

Provided that the adsorbing gas penetrates no further than the surfaces of the metals, μ_e will not be affected and the first two terms on the right-hand side of (2.11) will be zero. If, furthermore, the system can be chosen or arranged so that the gas adsorbs on to one metal only, say α, then $\chi_2^\beta = \chi_1^\beta$ and (2.11) becomes

$$\Delta^{\alpha\beta}\psi_2 - \Delta^{\alpha\beta}\psi_1 = \Delta V = \chi_2^\alpha - \chi_1^\alpha, \qquad \blacktriangleleft\textbf{(2.12)}$$

where ΔV is known as the *surface Volta potential difference*. Thus, two measurements of the Volta potential difference may be used to find the influence of an adsorbed film on the surface potential of a metal.

Volta potential differences between metals and aqueous solutions are also readily measurable and may be used in a precisely analogous way to indicate the effect of films on the surface potential of the aqueous solution. The argument in this instance involves the consideration of the electrochemical potentials of ions as well as of electrons, but the final result is still an equation similar to (2.12).

It must be remembered that if the film-forming substance is present to an appreciable extent in the solution, as would normally be so if the film were formed by adsorption rather than by spreading, the nature of the bulk phase may to some extent be affected and hence it may not be permissible to equate $(\mu)_2$ and $(\mu)_1$.

2.3. The measurement of Volta potential differences. Volta potential differences are the chief source of information regarding the electrical structure of the surfaces of condensed phases in contact with their vapour. The principles involved in the measurement of Volta potentials are the same regardless of the system, although there are various techniques which differ in detail.

The essentials are illustrated in fig. 2.1. Suppose that α and β are metals. One piece of β is connected to α and the other is placed a short

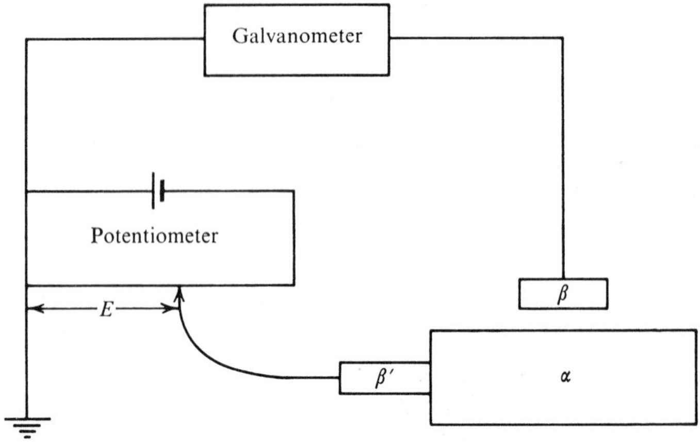

Fig. 2.1. The measurement of Volta (or compensation) potential differences.

distance from the surface of α. The potentiometer is adjusted until no current flows through the galvanometer. Under these conditions the following relationships hold. The outer potentials of α and β must be equal, as no current flows across the gap between them, i.e.

$$\psi^{\alpha} = \psi^{\beta}. \tag{2.13}$$

For the same reason the electrochemical potentials of the electrons in α and β' must be equal, and the electrochemical potentials of the electrons in β and β' must differ by Ee^-; i.e.

$$\tilde{\mu}_e^{\alpha} = \tilde{\mu}_e^{\beta'} \tag{2.14}$$

and

$$Ee^- = \tilde{\mu}_e^{\beta} - \tilde{\mu}_e^{\beta'}. \tag{2.15}$$

It is a simple exercise to show from (2.13) and (2.15) that provided β and β' are pieces of the same material (when $\alpha_e^{\beta} = \alpha_e^{\beta'}$), E is equal to the Volta potential difference between α and β'; i.e.

$$E = \Delta^{\alpha\beta'}\psi. \tag{2.16}$$

If a gas or vapour is permitted to adsorb on to the surface of α, the Volta potential difference changes, and the potentiometer must be adjusted to give zero current. The difference between the new (E_2) and original (E_1) readings of the potentiometer now gives the change in the surface potential of α. Thus from (2.12)

$$E_2 - E_1 = \Delta^{\alpha\beta'}\psi_2 - \Delta^{\alpha\beta'}\psi_1 = \Delta V = \chi_2^{\alpha} - \chi_1^{\alpha}. \qquad \blacktriangleleft\textbf{(2.17)}$$

If β and β' are not pieces of the same material, E is a measure of the Volta potential difference between α and β', together with a term which depends on the real potentials. Thus, although E is no longer equal to the Volta potential difference between α and β', when two measurements of E are made, as in surface potential studies, the term containing the real potentials subtracts out. True surface potential changes are consequently still given simply by the difference in the two E values.†

This conclusion also holds if α is an aqueous solution. It is, of course, necessary that the electrode β' should be in reversible equilibrium with α. The detailed equations are a little more complicated because instead of considering electrons, as in (2.14), it is necessary to consider the equilibrium of the ions of the metal β'. The full argument is given by Parsons (1954).

There are several methods for the measurement of Volta potential difference. The two most commonly used are referred to as the *radio-active probe* and the *vibrating plate* techniques. In the first, the gap between α and β is ionised by γ-radiation, usually from a source on β. The gaseous ions so produced carry the current between α and β when the outer potentials of the two phases are different, and this current is detected by means of a galvanometer consisting of a sensitive valve electrometer. The radio-active probe technique is therefore essentially a direct current method. The vibrating plate technique, on the other hand, is an alternating current method and is illustrated schematically in fig. 2.2. The metal probe β takes the form of a flat plate parallel to the surface of α and is made to vibrate sinusoidally in a direction normal to the α-surface. The surfaces of α and β then constitute the plates of a capacitor whose capacitance is given by

$$C = \frac{\epsilon_r \epsilon_0 \mathcal{A}}{d}.$$

The area, $\mathcal{A}$, of the plates and the dielectric constant, ϵ_r, of the space between them are constants, but the distance, d, between the plates and therefore the capacitance, C, vary with time in a manner determined by the sinusoidal vibrations. If there is a difference between the outer potentials of α and β an alternating current is generated. This is ampli-

† It is strictly correct to equate $(E_2 - E_1)$ and $(\Delta^{\alpha\beta'}\psi_2 - \Delta^{\alpha\beta'}\psi_1)$ only if there is evidence that the potentials applied from the potentiometer to establish zero current do not appreciably perturb the system. While for some systems such evidence exists, exceptions have been reported and it should be remembered that $(E_2 - E_1)$ is really a 'compensation' potential. Indeed, it is often referred to in these terms.

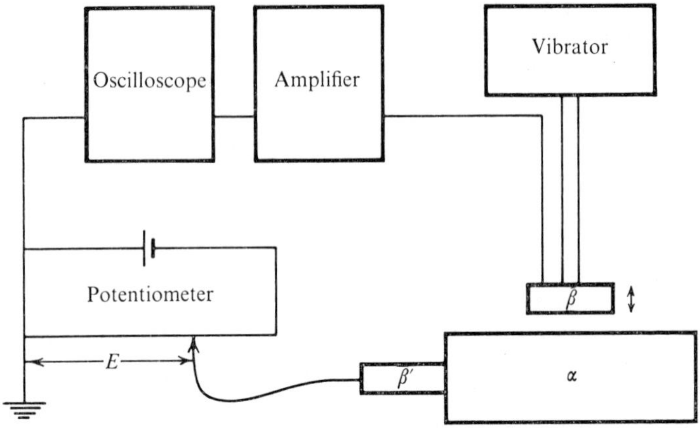

Fig. 2.2. A schematic diagram of the vibrating plate surface potentiometer.

fied and then detected by means of the oscilloscope. The procedure is then to adjust the potentiometer until the difference between the outer potentials, and hence the alternating current, becomes zero. Under these conditions, the potentiometer reading E obviously has the same significance as for the radio-active probe technique. It has been shown in studies of surface potentials at the air–water interface that the two techniques give similar results.

2.4. Surface potentials and the structure of interfaces. A great deal of attention has been given to the interpretation of changes in surface potential (χ) in terms of the arrangement of ions and molecules in surface phases. It has been stated earlier that χ originates from the presence of oriented dipoles of all kinds in a surface. It is now necessary, in order to describe the molecular theories of potentials at interfaces, to consider two different types of dipole. The first is simply a polarised atom or molecule, and the second is that formed by the segregation of ions of opposite charge (or of electrons and ions) in a direction normal to the surface. The contributions from the two types of dipole are often referred to as the *χ-potential*, in a restricted sense, and the *ionic double layer* potential, respectively. This terminology is not very satisfactory and the use of the term χ-potential in this sense will be avoided. For the surface of a phase in contact with its vapour

$$\chi^{\alpha} = \chi^{\alpha}(\text{molecular dipoles}) + \chi^{\alpha}(\text{ionic double layer}). \quad \blacktriangleleft \textbf{(2.18)}$$

For the interface between two condensed phases the difficulty of defining a meaningful potential analogous to χ has already been mentioned. Nevertheless the inner potential must change continuously from one phase to the other and may be interpreted in terms of molecular models which embody contributions from molecular dipoles and ionic double layers. Thus it is possible to write

$$\Delta^{\alpha\beta}\varphi = \Delta^{\alpha\beta}\varphi(\text{molecular dipoles}) + \Delta^{\alpha\beta}\varphi(\text{ionic double layer}). \quad \blacktriangleleft\mathbf{(2.19)}$$

The dipole and double layer terms will be discussed separately.

2.5. Molecular dipoles. The dipole terms in (2.18) and (2.19) may be related to molecular parameters of the surface in the following way. An array of surface dipoles is equivalent to the two plates of a charged capacitor. If the two plates are regarded as being separated by a distance equal to that between the two poles of the dipoles (measured normal to the surface) the potential difference between the plates is given by electrostatic theory as

$$\chi^{\alpha}(\text{molecular dipoles}) = \frac{Np^{s}}{\epsilon_0}. \quad \blacktriangleleft\mathbf{(2.20)}$$

In this equation N is the number of dipoles per unit area of surface, p^s is the normal component of each dipole moment and ϵ_0 is the permittivity of free space. According to the sign convention mentioned earlier, χ^{α} will be positive if the positive ends of the dipoles are towards the inside of the phase α. The magnitude and sign of the dipole potential as given by (2.20) can, at best, only be inferred from various indirect considerations and is not known with certainty for any system.

Changes in the molecular dipole potential can be measured *provided the ionic double layer potential remains constant* (see (2.18)). Thus under such conditions (2.17), (2.18) and (2.20) give

$$\Delta V = \frac{\Delta(Np^{s})}{\epsilon_0}. \quad \blacktriangleleft\mathbf{(2.21)}$$

Equation (2.21) has frequently been used to interpret the ΔV potentials which are produced by the spreading of insoluble monolayers at air–water interfaces (see §3.7). Whatever the situation in particular systems it must be emphasised that, in general, both ionic double layer and molecular dipole potentials will contribute to the total potential drop across an interface and it is not permissible to assume that either potential can be varied independently of the other.

2.6. The ionic double layer: Gouy–Chapman theory and the diffuse double layer. When two phases are in contact the phase boundary often carries a net positive or negative charge, which may result from the presence of electrons (e.g. when one phase is a metal) or of ions. At a solid–liquid interface, for example, a net ionic charge may arise either from the preferential dissolution of one of the ions of the solid lattice or through the preferential adsorption of one ionic species from the liquid phase. At a liquid–vapour interface a net charge may also arise from the spreading of an ionic substance which consists of one insoluble and one soluble ion. In systems such as these, soluble ions are retained in the surface phase adjacent to the phase boundary by the electrostatic attraction of the boundary charge.

The ionic double layer so formed has been treated theoretically by numerous authors at various levels of detail. Most treatments, however, rest ultimately on the theories of Gouy and Chapman and of Stern. These two theories are complementary. As will be seen, Stern modified the Gouy–Chapman model in order to bring the latter more into line with physical reality.

Gouy (1910) and Chapman (1913) proposed independently the so-called *diffuse double layer* theory of plane interfaces. The principles underlying the theory are very similar to those underlying the theory of strong electrolytes subsequently proposed by Debye and Hückel in 1923. Mathematically the Gouy–Chapman theory is the simpler because the equations are developed for planar rather than for spherical symmetry.

Consider an infinite plane surface carrying a non-discrete or smeared surface charge, in contact with a solution containing ions which may be regarded as point charges. At an infinite distance from the surface the electrical potential will be equal to the inner potential of the solution but, as the surface is approached, the potential gradually changes. It will be assumed that the work required to bring an ion from the bulk of the solution to a point near the surface, is entirely due to the electrical potential difference between the initial and final positions. The distribution of ions in the solution in a direction normal to the surface is then given by the following form of the Boltzmann equation,

$$N_i(x) = N_i(\infty)\exp[-z_i e^-(\varphi(x)-\varphi(\infty))/kT], \qquad (2.22)\dagger$$

† In many treatments in the literature $\varphi(\infty)$ has been taken arbitrarily as zero and $\varphi(x)-\varphi(\infty)$ has been abbreviated to $\varphi(x)$ (or sometimes $\psi(x)$). Thus, in such treatments all potentials are considered implicitly as relative to the bulk of one of the phases adjacent to the interface.

where $N_i(x)$ and $N_i(\infty)$ are respectively the numbers of ions per unit volume of species i, of valence z_i, at a distance x from the surface and at an infinite distance from the surface.

The relationship between the electrical potential and the space charge density $\rho(x)$, i.e. the net charge per unit volume at a point, is given by the Poisson equation which, for variations in the x direction only, takes the form

$$\frac{\partial^2 \varphi(x)}{\partial x^2} = -\frac{\rho(x)}{\epsilon_r \epsilon_0}, \tag{2.23}$$

where ϵ_r, the dielectric constant of the solution, is assumed not to vary with distance from the surface. The quantity $\rho(x)$ is given by

$$\rho(x) = \sum_i z_i e^- N_i(x) \tag{2.24}$$

and therefore from (2.22), (2.23) and (2.24),

$$\frac{\partial^2 \varphi(x)}{\partial x^2} = -\frac{1}{\epsilon_r \epsilon_0} \sum_i z_i e^- N_i(\infty) \exp\left[-z_i e^-(\varphi(x) - \varphi(\infty))/kT\right]. \tag{2.25}$$

A first integration of (2.25) may be carried out by use of the identity

$$2\frac{\partial^2 \varphi}{\partial x^2} = \frac{\partial}{\partial x}\left(\frac{\partial \varphi}{\partial x}\right)^2.$$

Thus, if each side of (2.25) is multiplied by $2(\partial\varphi(x)/\partial x)$ the equation may be integrated to give

$$\left(\frac{\partial \varphi(x)}{\partial x}\right)^2 = \frac{2kT}{\epsilon_r \epsilon_0} \sum_i N_i(\infty) \{\exp\left[-z_i e^-(\varphi(x) - \varphi(\infty))/kT\right] - 1\}, \tag{2.26}$$

where the boundary conditions $x \to \infty$, $\varphi(x) \to \varphi(\infty)$ and $\partial\varphi(x)/\partial x \to 0$ have been used.

The double layer as a whole is electrically neutral and, in consequence, the charge on the surface must be balanced by the charge in the solution; i.e.

$$\sigma = -\int_0^\infty \rho(x)\,\mathrm{d}x, \tag{2.27}$$

where σ is the two-dimensional density of charge on the surface. On substitution for $\rho(x)$ from (2.23), (2.27) becomes

$$\sigma = \int_0^\infty \epsilon_r \epsilon_0 \frac{\partial^2 \varphi(x)}{\partial x^2}\,\mathrm{d}x,$$

which, with the boundary conditions given above, can be integrated to yield

$$\sigma = -\epsilon_r \epsilon_0 \left(\frac{\partial \varphi(x)}{\partial x}\right)_{x=0}. \tag{2.28}$$

Combination of (2.25) and (2.28) then gives

$$\sigma = [2\epsilon_r \epsilon_0 kT \sum_i N_i(\infty) \{\exp[-z_i e^-(\varphi(0)-\varphi(\infty))/kT]-1\}]^{\frac{1}{2}}, \quad \blacktriangleleft\mathbf{(2.29)}$$

where $\varphi(0)$ is the potential at the surface.† The negative root of (2.26) has been chosen because when σ is positive φ must become less positive with distance from the surface; i.e. $(\partial\varphi(x)/\partial x)_{x=0}$ must be negative. Equation (2.29) is commonly referred to as the Gouy–Chapman equation.

Equations (2.26) and (2.29) are cumbersome to use and become considerably simpler if it is supposed that the system contains only one symmetrical electrolyte. In this instance (2.26) may be written

$$\frac{\partial \varphi(x)}{\partial x} = -\left(\frac{8N(\infty)\,kT}{\epsilon_r \epsilon_0}\right)^{\frac{1}{2}} \sinh \frac{ze^-}{2kT}(\varphi(x)-\varphi(\infty)) \tag{2.30}$$

and (2.29) may similarly be reduced to

$$\sigma = (8N(\infty) e_r \epsilon_0 kT)^{\frac{1}{2}} \sinh \frac{ze^-}{2kT}(\varphi(0)-\varphi(\infty)), \quad \blacktriangleleft\mathbf{(2.31)}$$

where $N_+(\infty) = N_-(\infty) = N(\infty)$ and $z_+ = -z_- = z$.

On further integration (2.30) yields

$$\kappa x = \ln\left\{\frac{[\exp\{ze^-(\varphi(x)-\varphi(\infty))/2kT\}+1]\,[\exp\{ze^-(\varphi(0)-\varphi(\infty))/2kT\}-1]}{[\exp\{ze^-(\varphi(x)-\varphi(\infty))/2kT\}-1]\,[\exp\{ze^-(\varphi(0)-\varphi(\infty))/2kT\}+1]}\right\}, \tag{2.32}$$

where κ is identical to the recriprocal length parameter of Debye–Hückel theory; i.e. for a symmetrical electrolyte

$$\kappa = \left(\frac{2z^2(e^-)^2 N(\infty)}{\epsilon_r \epsilon_0 kT}\right)^{\frac{1}{2}}. \tag{2.33}$$

† It will be remembered that in this discussion the contributions to the Galvani potential difference of oriented molecular dipoles and of the ionic double layer are being considered separately. The dipoles were dealt with in the preceding section and here only the ionic double layer is treated. For this reason, the term $(\varphi(0)-\varphi(\infty))$ is only the contribution to the potential from the ionic double layer, although there will almost certainly be a dipole potential difference between $x = 0$ and $x = \infty$. The same artificial separation of effects occurs again in the treatment of the Stern theory below.

Even simpler forms of the diffuse double layer equations may be obtained from (2.30), (2.31) and (2.32) if $ze^-(\varphi(x)-\varphi(\infty)) \ll kT$. Under these conditions it is permissible to expand the exponential terms and neglect all but the first and second order terms in the resulting series. Expression (2.30) then becomes

$$\frac{\partial\varphi(x)}{\partial x} = -\left(\frac{8N(\infty)kT}{\epsilon_r\epsilon_0}\right)^{\frac{1}{2}} \frac{ze^-}{2kT}(\varphi(x)-\varphi(\infty)) \qquad (2.34)$$

while (2.31) simplifies to

$$\sigma = \epsilon_r\epsilon_0\kappa(\varphi(0)-\varphi(\infty)) \qquad \blacktriangleleft\mathbf{(2.35)}$$

and (2.32) becomes

$$\varphi(x)-\varphi(\infty) = (\varphi(0)-\varphi(\infty))\exp(-\kappa x). \qquad \blacktriangleleft\mathbf{(2.36)}$$

Equations (2.31) and (2.35) suggest that the potential at the surface, $\varphi(0)$, is determined principally by the surface charge density and the bulk electrolyte concentration (figs. 2.3 and 2.4). However, while the potential increases with surface charge density (linearly at low potentials) it decreases with the square root of the electrolyte concentration. The decay of potential from the surface towards the bulk value $\varphi(\infty)$, is approximately exponential. At low potentials this is accurately so according to (2.36) and it can also be seen from this equation that the potential decays to $[\varphi(0)-\varphi(\infty)]/e$ at a distance $1/\kappa$ from the surface. It is customary to refer to $1/\kappa$ as the 'thickness' of the diffuse double layer. Fig. 2.4 shows potentials at various electrolyte concentrations for a fixed surface charge density which corresponds to an area of 10 nm^2 (1000 $Å^2$) per surface ion. Ionic concentrations at various distances from the surface are predicted by the Boltzmann equation (2.22). For a surface bearing a charge density of $+0.0158$ C m^{-2} ($(\varphi(0)-\varphi(\infty)) =$ 56.4 mV) the calculated concentration profiles are shown in fig. 2.5. It should be noted that the anions are attracted to, and the cations repelled from the surface.

The application of the Gouy–Chapman diffuse double layer theory to actual systems will be discussed in later chapters, but it is appropriate to mention here three of its more important shortcomings.

(i) The dielectric constant has been assumed to be independent of distance from the surface. Now the dielectric constant of a dipolar liquid such as water is known to vary with the electric field strength, and from (2.30) it can be deduced that in a diffuse double layer, field strengths of more than 10^8 V m^{-1} may be reached. There is evidence that in field strengths such as these the dielectric constant of water may

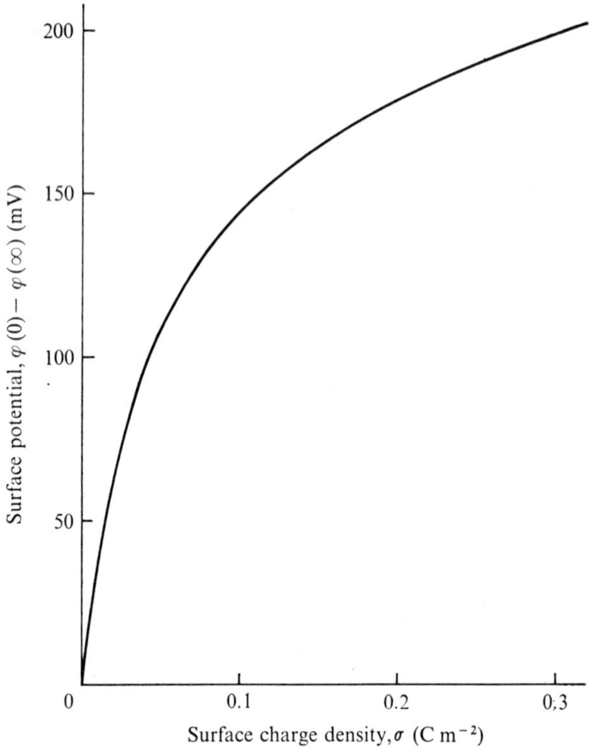

Fig. 2.3. The surface potential as a function of the surface charge density, according to the Gouy–Chapman equation (2.29). The curve is for a uni-univalent electrolyte at 10^{-2} mol l^{-1}. The dielectric constant has been taken as 80 and the temperature as 20 °C.

fall by an order of magnitude. The assumption of constant dielectric constant is therefore of doubtful validity, although the errors entailed are thought not to be large.

(ii) All ions have been assumed to be point charges. This assumption is particularly unsatisfactory in two respects. Firstly the number of point ions that can be concentrated into unit volume is infinite, while for actual ions this number is strictly limited by their volume. It is therefore possible, and can easily happen, that (2.22) predicts absurdly high local ion concentrations for quite reasonable assumptions as to the potential and bulk electrolyte concentration. Secondly, as point ions may approach infinitely closely to the surface, $\varphi(0)$ is necessarily the potential

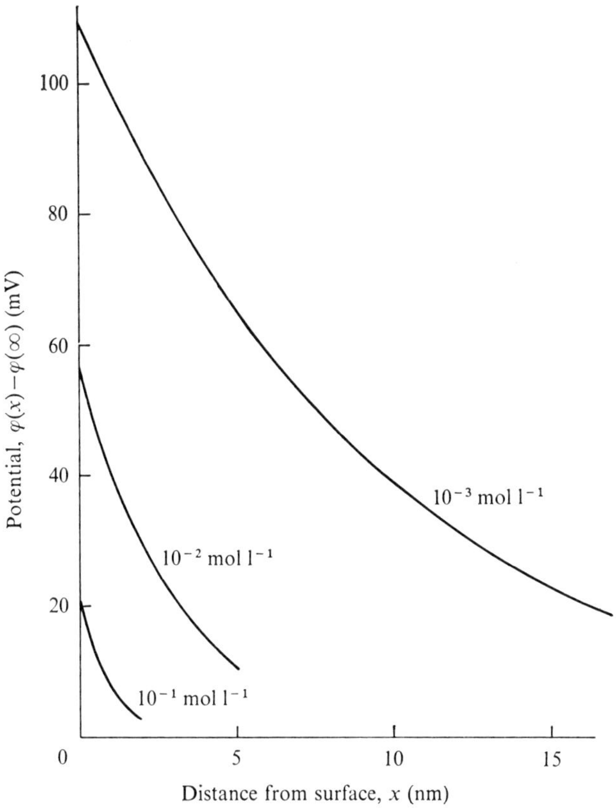

Fig. 2.4. The decay of potential from a surface according to (2.32). The surface charge density has been assumed to be 0.0158 C m^{-2}. The electrolyte is uni-univalent and the dielectric constant and temperature have been taken as 80 and 20 °C respectively.

at the surface. For real ions, however, the distance of closest approach to the surface is equal to the effective radius of the ion. The lower limit of the integral in (2.27) should then be a function of the effective ionic radius, say δ, instead of zero and the potential $\varphi(0)$ should be replaced in double layer equations by $\varphi(\delta)$. This correction is incorporated in the Stern theory, which will be discussed below.

(iii) The surface charge has been assumed to be smeared over the surface rather than, as it actually is, in the form of discrete ions and electrons. The diffuse layer in reality consists of the overlapping ionic atmosphere of each individual surface charge, and the potential in a

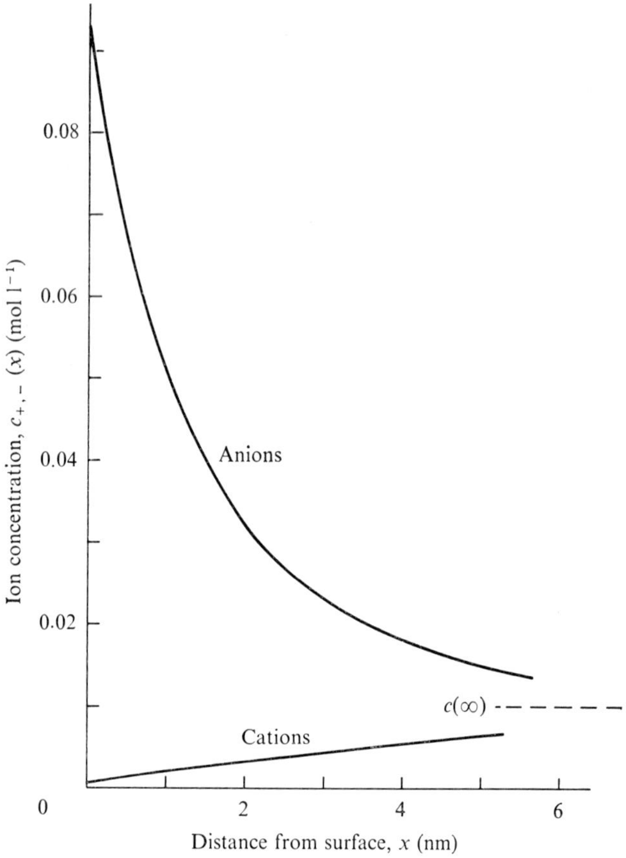

Fig. 2.5. The concentrations of univalent cations and anions as a function of distance from a charged surface according to (2.22) and (2.32). $\varphi(0)-\varphi(\infty)$ has been taken as 56.4 mV and the bulk electrolyte concentration $c(\infty)$ as 10^{-2} mol l^{-1}. The dielectric constant and temperature have been assumed to be 80 and 20 °C respectively.

plane parallel to the surface fluctuates from place to place according to the degree of overlap of these atmospheres. The potentials in the Gouy–Chapman theory are thus average potentials. As far as the properties of the diffuse layer are concerned this averaging probably does not introduce much error, but for the specific adsorption of ions, as in the Stern theory, the assumption of smeared charge is thought to be less valid.

Other possible corrections to the Gouy–Chapman theory have been examined from time to time and the interested reader is referred, as a starting point, to Levine & Bell (1966).

2.7. The ionic double layer: Stern theory and the molecular capacitor. The theory of Stern (1924) extends the preceding double layer theory by the introduction of two modifications. One of these, as mentioned in (ii) above, is essentially a correction and takes into account the fact that ions of finite size cannot approach more closely to the surface than a distance equal to their effective radius. But the other is a new concept which recognises that at short distances from the surface there may exist a specific 'chemical' interaction between the ions and the surface or, in other words, that the Boltzmann equation (2.22) may contain work terms other than that arising from the electrical potential difference. The following description of Stern's theory is not as it was originally presented. It has been modified so as to show how the theory follows from the principles discussed in chapter 1.

Fig. 2.6 shows the total surface charge σ (assumed to be positive) balanced by the charge σ_{st} of ions in the Stern layer, whose centres are at distance δ from the surface, and the charge σ_d of ions of the diffuse layer. Thus

$$\sigma = -(\sigma_{st}+\sigma_d). \tag{2.37}$$

The problem is to find an expression for σ_{st}.

If $(N_1)_i$ is the number of ions of species i adsorbed per unit area of the Stern layer, then

$$\sigma_{st} = \sum_i z_i e^-(N_1)_i. \tag{2.38}$$

The calculation of N_1 from a theoretical adsorption isotherm has now to be attempted. The treatment used by Stern is similar in principle to that used by Langmuir for the adsorption of gases, and rests on the assumption of a homogeneous surface, localised adsorption and no interaction (other than coulombic) between adsorbed molecules. Equation (1.87) does not exactly fit the present requirements but can readily be modified to do so. For instance, for the simultaneous adsorption of two gases, rather than one, it is easy to show that (1.87) can be written in the form

$$\frac{\theta_i}{1-\theta_i-\theta_j} = K_i' p_i \tag{2.39}$$

and that there is a corresponding equation for the adsorption of j. p_i is the partial pressure of i. For a solution of an electrolyte in which

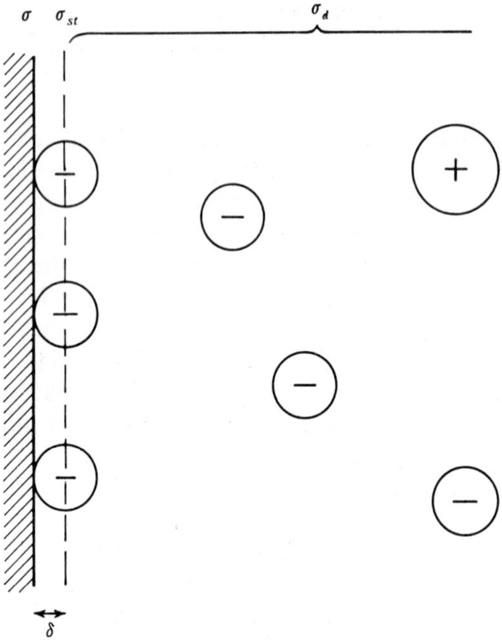

Fig. 2.6. The electrical double layer according to Stern.

the solvent is effectively a structureless dielectric, *i* and *j* can be thought of as the positive and negative ions and (2.39) may be transformed into

$$\frac{\theta_+}{1-\theta_+-\theta_-} = K_+ x_+ \tag{2.40}$$

where x_+ is the mole fraction of the positive ion and K and K' are related through Henry's law. In (2.40),

$$\left.\begin{aligned} \theta_+ &= \frac{(N_1)_+}{N_s} \\ \text{and} \qquad \theta_- &= \frac{(N_1)_-}{N_s} \end{aligned}\right\} \tag{2.41}$$

where N_s is the number of adsorption sites per unit area of surface and K, instead of being a function of standard chemical potentials, is now a function of standard electrochemical potentials. Thus, by reference to (1.95) it can be seen that

$$\left.\begin{aligned} K_+ &= \exp\{-[(\tilde{\mu}_+^{\ominus})_{x=\delta_+} - (\tilde{\mu}_+^{\ominus})_{x=\infty}]/kT\} = \exp\{-\Delta_a\tilde{\mu}_+^{\ominus}/kT\}, \\ K_- &= \exp\{-[(\tilde{\mu}_-^{\ominus})_{x=\delta_-} - (\tilde{\mu}_-^{\ominus})_{x=\infty}]/kT\} = \exp\{-\Delta_a\tilde{\mu}_-^{\ominus}/kT\}. \end{aligned}\right\} \tag{2.42}$$

It is further assumed that $\Delta_a\tilde{\mu}^{\ominus}_{+,-}$ may be usefully split into its 'chemical' and 'electrical' components such that

$$\Delta_a\tilde{\mu}^{\ominus}_{+,-} = \Delta_a\mu^{\ominus}_{+,-} + z_{+,-}e^-(\varphi(\delta)-\varphi(\infty)). \tag{2.43}$$

From (2.38), (2.40) and (2.41) the desired equation for σ_{st} may be obtained. For simplicity it is assumed that $z_+ = -z_- = z$, and $x_+ = x_- = x$. Then,

$$\sigma_{st} = \frac{ze^-N_s x(K_+ - K_-)}{1+x(K_+ + K_-)}. \tag{2.44}$$

This equation was originally derived by Esin & Markov (1939). In practice it is usually found that one ion is adsorbed at the δ-plane much more strongly than the other and, in consequence, either K_+ or K_- may be neglected. If, for instance, the cations are very weakly adsorbed, (2.44) combined with (2.42) and (2.43) reduces to

$$\sigma_{st} = \frac{-ze^-N_s x \exp[-\Delta_a\mu^{\ominus}_-/kT]\exp[ze^-(\varphi(\delta)-\varphi(\infty))/kT]}{1+x\exp[-\Delta_a\mu^{\ominus}_-/kT]\exp[ze^-(\varphi(\delta)-\varphi(\infty))/kT]} \quad \blacktriangleleft\textbf{(2.45)}$$

and, furthermore, if the solution is dilute (x is small) it may be possible to neglect the second term in the denominator of (2.45) in comparison with unity, and so obtain the even simpler equation

$$\sigma_{st} = -ze^-N_s x\exp[-\Delta_a\mu^{\ominus}_-/kT]\exp[ze^-(\varphi(\delta)-\varphi(\infty))/kT]. \tag{2.46}$$

In (2.45) and (2.46) $(\varphi(\delta)-\varphi(\infty))$ is given by the diffuse layer theory as (see 2.31)

$$\sigma_d = (8N(\infty)\epsilon_r\epsilon_0 kT)^{\frac{1}{2}}\sinh\frac{ze^-}{2kT}(\varphi(\delta)-\varphi(\infty)). \tag{2.47}$$

The relationship of σ_{st} to the total charge on the surface has already been given in (2.37). Equation (2.45) is often referred to as the Stern equation.

A further important relationship is that between the capacitances C_{st} and C_d of the Stern and diffuse layers respectively, and that of the total double layer, C. As the layers are in series with each other

$$C = \frac{C_{st}C_d}{C_{st}+C_d}, \quad \textbf{(2.48)}$$

where
$$C_{st} = \frac{\sigma}{\varphi(0)-\varphi(\delta)} \quad \blacktriangleleft\textbf{(2.49)}$$

and
$$C_d = \frac{\sigma}{\varphi(\delta)-\varphi(\infty)}. \quad \blacktriangleleft\textbf{(2.50)}$$

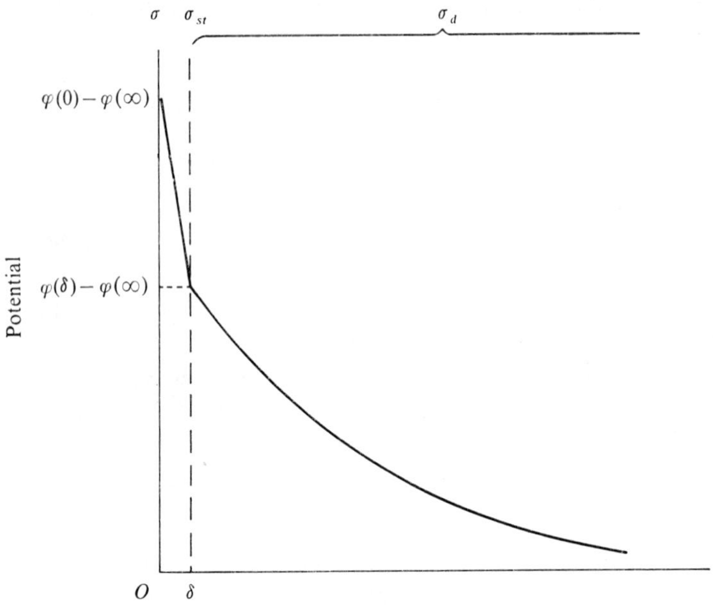

Fig. 2.7. The decay of potential with distance from the surface according to Stern.

The total capacitance C should be determined by both C_{st} and C_d, but under extreme conditions of electrolyte concentration it should tend to either one or the other. At very low concentrations,

$$\sigma_{st} \to 0, \quad \varphi(\delta) \to \varphi(0), \quad C_{st} \to \infty \quad \text{and} \quad C \to C_d \tag{2.50a}$$

and the double layer capacitance becomes effectively that of the diffuse layer. The concentration at which this should become true obviously depends on the nature of the ions, and, in particular, on $\Delta_a \mu^{\ominus}$. At very high concentrations on the other hand,

$$\varphi(\delta) - \varphi(\infty) \to 0, \quad C_d \to \infty \quad \text{and} \quad C \to C_{st} \tag{2.50b}$$

and it is the properties of the Stern layer, or molecular capacitor, which should determine the total capacitance. The distribution of potential in an electrical double layer of the Gouy–Chapman–Stern type is usually assumed to be as in fig. 2.7, although variations such as those depicted in fig. 4.10 are possible. The form of the curve for $\delta \leqslant x \leqslant \infty$ is determined by (2.32), in which $\varphi(0)$ is replaced by $\varphi(\delta)$. For $0 \leqslant x \leqslant \delta$ the potential gradient is depicted as constant. In so much as there is no net charge (by definition) in this region this may be correct, but according to electrostatic theory it is also necessary that the dielectric constant of the medium should remain constant. In view of the close proximity of the surface and its possible influence on the orientation of the solvent molecules this assumption is rather dubious. It certainly seems probable that the potential drop from $x = 0$ to $x = \delta$ is due partly to oriented solvent dipoles (as discussed earlier in this section) as well as to the ionic charge, and there is good evidence that the dielectric constant in the Stern layer is well below that of the bulk solution.

The Gouy–Chapman and Stern theories will be discussed again in connection with their application to various systems (chapters 3 and 4). However, two further points may be mentioned here. Grahame (1947) has attempted to modify the above treatment by taking into account that in the Stern layer there may be two types of ion, one of which is specifically adsorbed, the other being merely a normal hydrated ion at its distance of closest approach to the surface. In Grahame's model the specifically adsorbed ion is assumed to lose some of its water of hydration and to be closer to the surface than the other ion. The concept of the single layer of ions with their centres all in one plane, as in fig. 2.6, must then be replaced by a model which consists of two parallel planes at different potentials, one associated with each type of ion. These two layers are usually called the *inner* and *outer Helmholtz layers*. The second point concerns discreteness of charge. It has been mentioned earlier that this effect is probably more important in the Stern than in the Gouy–Chapman or diffuse region of the double layer. The reason for this lies in the influence that it may have on the potential in the δ-plane. The ions in the Stern layer are assumed to adsorb on to lattice points which are empty prior to the arrival of the ion. The electrical work involved in the adsorption is therefore determined by the potential at the lattice point and not by the average potential in the δ-plane. Calculations have shown that these two potentials may differ significantly and that the discreteness of charge should be considered in the theory of ion adsorption.

2.8. The application and testing of double layer theory. The Gouy–Chapman–Stern theory provides a basis for the interpretation of the ionic double layer potential, in (2.18) and (2.19). If, for example, (2.20) is combined with (2.18) and substituted into (2.17), and also χ (double layer) is replaced by $\varphi(0)-\varphi(\infty)$

$$\Delta V = \frac{1}{\epsilon_0}[(Np^s)_2-(Np^s)_1]+[(\varphi(0)-\varphi(\infty))_2-(\varphi(0)-\varphi(\infty))_1]$$

or

$$\Delta V = \frac{\Delta(Np^s)}{\epsilon_0}+\Delta(\varphi(0)-\varphi(\infty)). \quad \blacktriangleleft\textbf{(2.51)}$$

This equation has been used extensively in attempts to interpret Volta (compensation) potential measurements at air–water and hydrocarbon–water interfaces. The ionic double layer term on the right-hand side of (2.51) is calculated either from diffuse layer theory or from the combined diffuse and Stern layer theory.

It is doubtful whether there are any experimental systems to which the double layer theory set out above can be applied, or in which it can be tested, without some degree of ambiguity. This stems ultimately from the arbitrary splitting of the electrochemical potential into chemical and electrical components (§2.2). By far the most amenable system is that of mercury in contact with an aqueous solution (chapter 4). In general, however, electrical double layers at interfaces are structurally very complicated and the measurable quantities very few. The principles which govern the behaviour of systems containing double layers, as expounded in the above theory, have nevertheless contributed greatly to the understanding of surface phenomena.

2.9. Electrokinetic potentials. In the following short description of electrokinetic potentials the aim is not to provide in any sense a complete treatment, but rather to emphasise the relationship (or lack of it!) between these potentials and those discussed so far in this chapter.

Consider a plane surface of an insulating material in contact with an electrolyte solution, such that there is a Gouy–Chapman diffuse double layer at the interface, and a uniform electric field $\boldsymbol{E}$ in a direction parallel to the surface (fig. 2.8). The ions of the diffuse layer will experience a force parallel to the surface due to the applied field and will transmit this force through friction to the surrounding solvent. For an element of solution of unit area and thickness dx this force will be equal to $\boldsymbol{E}\rho\,dx$ where, as previously, ρ is the space charge density. In the steady state this force will be exactly balanced by the viscous drag of the solution on either side of the element, and the equation

$$\boldsymbol{E}\rho(x)\,dx = \left(\eta\frac{\partial v(x)}{\partial x}\right)_{x+dx} - \left(\eta\frac{\partial v(x)}{\partial x}\right)_{x} \qquad (2.52)$$

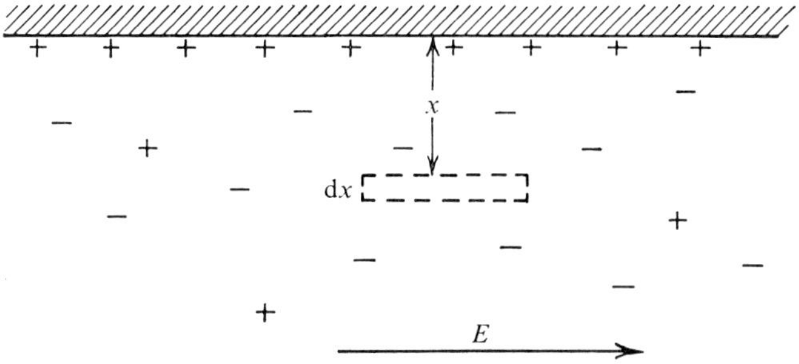

Fig. 2.8. The point-charge model of the electrical double layer assumed for the derivation of the electrokinetic equations.

holds, where v is the velocity of the solution parallel to the surface and η is its viscosity. If it is assumed that the viscosity is independent of distance from the surface,

$$\boldsymbol{E}\rho(x)\,\mathrm{d}x = \eta\frac{\partial^2 v(x)}{\partial x^2}\mathrm{d}x. \tag{2.53}$$

On substituting for ρ from the Poisson equation (2.23), (2.53) becomes

$$-\epsilon_r\epsilon_0\boldsymbol{E}\frac{\partial^2\varphi(x)}{\partial x^2}\mathrm{d}x = \eta\frac{\partial^2 v(x)}{\partial x^2}\mathrm{d}x \tag{2.54}$$

and on integration

$$-\epsilon_r\epsilon_0\boldsymbol{E}\frac{\partial\varphi(x)}{\partial x} = \eta\frac{\partial v(x)}{\partial x}+\text{constant}. \tag{2.55}$$

The integration constant may be obtained from the boundary conditions

$$x\to\infty,\quad \partial\varphi(x)/\partial x\to 0,\quad \partial v(x)/\partial x\to 0$$

and is obviously zero. Further integration of (2.55) gives

$$-\epsilon_r\epsilon_0\boldsymbol{E}\varphi(x) = \eta v(x)+\text{constant}. \tag{2.56}$$

If it is assumed that there is no slippage of liquid at the surface, then this integration constant may be determined from the boundary condition

$$v = 0,\quad \varphi(x) = \zeta \tag{2.56a}$$

and therefore

$$-\epsilon_r\epsilon_0\boldsymbol{E}\varphi(x) = \eta v(x)-\epsilon_r\epsilon_0\boldsymbol{E}\zeta. \tag{2.57}$$

As $\quad x\to\infty,\quad \varphi(x)\to\varphi(\infty)\quad\text{and}\quad v(x)\to v_E,$

where v_E is the maximum velocity acquired by the liquid, and hence from (2.57)

$$\zeta - \varphi(\infty) = \frac{\eta}{\epsilon_r \epsilon_0 \boldsymbol{E}} v_E. \tag{2.58}$$

$\varphi(\infty)$ is usually taken arbitrarily as zero, in which case

$$\zeta = \frac{\eta}{\epsilon_r \epsilon_0 \boldsymbol{E}} v_E. \qquad \blacktriangleleft \ \textbf{(2.59)}$$

Equation (2.59) is known as the *Helmholtz–Smoluchowski* equation, ζ as the *electrokinetic* or *zeta potential* and v_E as the *electrokinetic velocity*. The term $v_E/\boldsymbol{E}$ is referred to as the *electrokinetic mobility*, u_E. ζ is defined only by the boundary condition (2.56*a*), and is therefore the potential in a plane in the liquid (known as the *plane of shear*) close to the surface. The physical significance of ζ and the precise location of the plane of shear are important questions and will be taken up again shortly.

Electrokinetic phenomena include four distinct processes. As it has been derived, (2.59) relates to the movement of liquid adjacent to a solid surface under the influence of an electric field parallel to the surface. This process is known as *electro-osmosis* and may be observed when an electrical potential difference is applied between the ends of a capillary tube or porous plug filled with an electrolyte solution.

Instead of considering the motion of the liquid relative to the solid surface, the reverse situation could have been considered. In this instance the process would be called *electrophoresis*, v_E would be the velocity of the solid surface, and u_E would be called the *electrophoretic mobility*. In order to demonstrate electrophoresis the solid has to be in the form of particles which are sufficiently small to remain suspended in the liquid during the application of the field and the measurement of their velocity. While all particles down to the dimensions of ions exhibit electrophoresis it must be remembered that (2.59) was derived for a plane surface and can be shown to be quantitatively correct only if the radius of curvature of the particle is much larger than the effective thickness ($1/\kappa$) of the electrical double layer at its surface.

The third electrokinetic process is that which gives rise to *streaming potentials*. As might be expected from (2.59), if liquid is forced to flow past a surface having a finite zeta potential, an electrical potential difference (the streaming potential) is set up between the ends of the surface in such a way that it tends, by electro-osmosis, to inhibit the flow. Streaming potentials are readily measurable when electrolyte solutions are forced through porous plugs.

The fourth process involves the movement of particles through a liquid under the influence of, for example, a gravitational field and gives rise to the migration or *sedimentation* potential. Unlike the streaming potential, the migration potential is difficult to measure and has been little studied.

The foregoing paragraphs give only an outline of the principles of electrokinetics as derived for an infinite plane surface. A more complete discussion, including the practical problems of measurement, would involve an excursion into colloid chemistry and is beyond the scope of this book. For further information on these matters the reader should refer to the articles by Overbeek (1950, 1952). The remainder of this section will be concerned only with the significance of electrokinetic potentials.

The electrokinetic velocity v_E is measurable and unambiguous. The electrokinetic or zeta potential, which is derived from the velocity by means of (2.59) is, however, defined only as the potential at the plane of shear of the liquid. In order to attach significance to the zeta potential and to relate it to the structure of the double layer, it is necessary to know the position of the plane of shear. Herein lies the principal difficulty in the interpretation of electrokinetic data. Probably the simplest approach is to assume that the counter-ions are point charges, and that the layer of solvent molecules in contact with the surface does not move. The latter part of this assumption is, of course, the usual boundary condition of no slippage. The plane of shear and the zeta potential would then be located at approximately the diameter of a solvent molecule from the surface. However, if the ions adjacent to the surface (in the Stern layer) are of finite size and effectively immobilised, then it is likely that the plane of shear would be further from the surface. The precise position of the plane of shear in this model would probably depend on the surface density of the ions in the Stern layer. Thus, if these ions were well separated, ions which were not adsorbed could well move almost unhindered parallel to the surface down to distances almost equal to their effective radii. The plane of shear would then be at a distance very close to the Stern (δ) plane (fig. 2.9). If, on the other hand, the ions adsorbed in the Stern layer were very close to each other the non-adsorbed ions might not be able to shear freely unless they were further out. In the limit of this situation the plane of shear might be at a distance closer to 3δ, or three ionic radii from the surface.

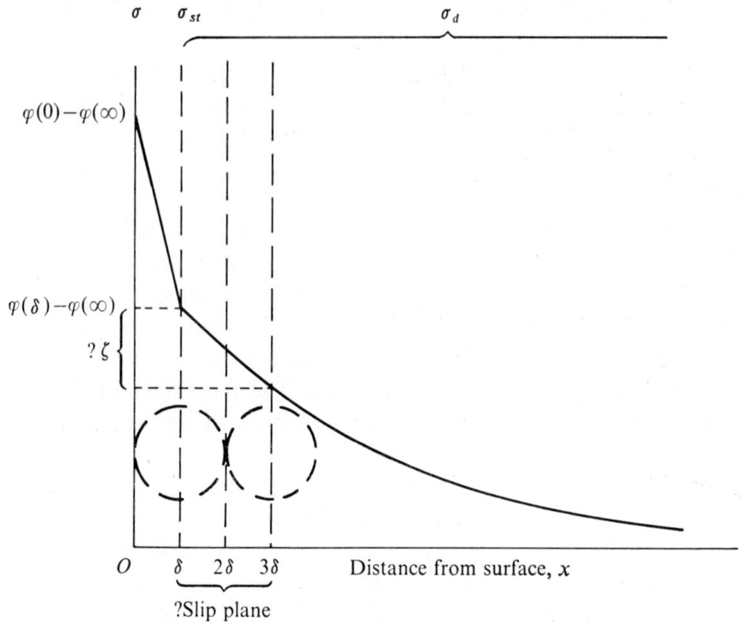

Fig. 2.9. An illustration of one of the factors which may affect the position of the plane of shear.

These considerations ignore so-called structural effects resulting from the possible long range influence of the solid on the visco-elastic properties of the solution. It has often been proposed that certain types of solid surface promote structure in the solution in such a way that shearing occurs only at distances of several or even many solvent diameters from the surface. Such effects are, however, not very well established.

Even if the existence of elasticity in the solvent is disregarded the question still arises as to the value of the dielectric constant and viscosity to be used in (2.59). Both these quantities relate to the ionic double layer (not the bulk solution) and may be influenced by the very high field strengths and ion concentrations close to the surface. In the discussion of the Gouy–Chapman theory it was commented that it seemed unlikely that the use of the bulk value of the dielectric constant would produce much error. The situation concerning the viscosity, however, is less clear, as there is no satisfactory theory and little or no direct experimental evidence as to its magnitude close to an interface. The most that can be said is that there is no incontrovertible evidence that the viscosity in aqueous diffuse double layers is greatly different to that in the bulk.

The difficulty of relating the electrokinetic potential to the double layer structure can be seen clearly from the uncertainty of the position of the plane of shear. If δ is taken to be 0.15 nm (1.5 Å), then according to fig. 2.9 the plane of shear might be located between 0.15 and 0.45 nm. In uni-univalent electrolyte at 0.1 mol l^{-1} the thickness of the diffuse layer ($1/\kappa$) is approximately 1 nm. With an uncertainty in position of 0.3 nm, particularly in the region of high potential gradient near the wall, the potential of the plane of shear cannot be deduced from the double layer structure (or vice versa) with any worthwhile precision. For low electrolyte concentrations the situation should improve, although the evidence that it does so is not impressive. It is to be hoped that future investigations will throw some further light on the problem of the location of the plane of shear. If it were not for this difficulty the techniques of electrokinetics would be very much more powerful tools in the examination of the ionic structure of surfaces.

References

Chapman, D. L. (1913). *Phil. Mag.* (6) **25**, 475.

Debye, P. & Hückel, E. (1923). *Physik. Z.* **24**, 185.

Esin, O. A. & Markov, B. F. (1939). *Acta physicochim.* **10**, 353.

Gibbs, J. W. (1928). *Collected Works*, 2nd edition (Longmans, Green, New York): vol. 1, p. 429.

Gouy, G. (1910). *J. Phys.* **9**, 457.

Grahame, D. C. (1947). *Chem. Revs.* **41**, 441.

Guggenheim, E. A. (1929). *J. Phys. Chem.* **33**, 842.

Guggenheim, E. A. (1967). *Thermodynamics*, 5th edition (North Holland Publishing Co., Amsterdam): p. 298.

Levine, S. & Bell, G. M. (1966). *Disc. Faraday Soc.* **42**, 69.

Overbeek, J. Th. G. (1950). In *Advances in Colloid Science* (ed. H. Mark & E. J. W. Verwey), (Interscience, New York and London): vol. 3, p. 97.

Overbeek, J. Th. G. (1952). In *Colloid Science* (ed. H. R. Kruyt), (Elsevier, Amsterdam): vol. 1, p. 130.

Parsons, R. (1954). In *Modern Aspects of Electrochemistry* (ed. J. O'M. Bockris & B. E. Conway), (Butterworths, London): p. 103.

Stern, O. (1924). *Z. Elektrochem*, **30**, 508.

3 Liquid interfaces

3.1. Introduction. This chapter is the first in which the surface chemistry of specific systems is discussed. Liquid interfaces have been chosen because they are usually well-defined and therefore, in principle, easier to treat than systems containing the less well-defined solid surfaces. Nevertheless the range of phenomena encountered in liquid systems is very extensive. Only a small proportion of the field is covered in the following sections and this consists necessarily of the more basic topics.

A very large part of the study of liquid interfaces centres round the measurement of surface or interfacial tension and this measurement is closely connected with the properties of curved liquid surfaces. These two topics are discussed in §3.2 and §3.3. Then follows a description of the surfaces of pure liquids and of spreading and adhesion in systems of two immiscible liquids. Next to be discussed are the surface films formed in two or three component systems. The problem here arises as to the most suitable order of presentation. Thus, two component systems may consist in the one extreme of completely miscible liquid mixtures, and in the other of an insoluble monomolecular film of one component on the surface of the other, and quite different treatments and even experimental techniques are required in each case. In between these two extremes lie all manner of solutions on the surfaces of which films are formed by adsorption. The adsorbed films may, however, consist either of ions or of neutral molecules and for this reason it is convenient to subdivide the discussion further.

In the order chosen, the first topic to be presented is that of the completely miscible liquid mixtures. Next come successively sections on insoluble monolayers and adsorption from dilute solutions, in both of which electrolytes and non-electrolytes are considered separately.

3.2. Curved interfaces: the Laplace and Kelvin equations. An understanding of phenomena at curved liquid surfaces is basic to the treatment of several processes of importance in surface chemistry. In the present context, a knowledge of the pressure drop across curved liquid surfaces is necessary for the experimental determination of the surface tension of liquids by the capillary rise method. The expression which relates this pressure drop to the principal radii of curvature of the surface and the surface tension of the liquid is called the Laplace equation. It is also known that the vapour pressure of a liquid depends on the curvature of its surface, as expressed by the Kelvin equation. This equation is of fundamental importance and, for example, yields a ready explanation of the condensation of vapours in small capillaries at a pressure below the saturated vapour pressure. This application is used in connection with the adsorption of gases on porous solids in §5.18.

Suppose that in a one component system a vapour bubble exists in equilibrium with liquid. In the absence of any external field, e.g. gravitational or electrical, the bubble will assume a spherical shape of radius of curvature r. In a hypothetical contraction of the bubble to a radius of $r-\mathrm{d}r$, the change in surface area would be $8\pi r\,\mathrm{d}r$ and thus, the decrease in surface free energy would be $8\pi r\,\mathrm{d}r\gamma$. The terms in $\mathrm{d}r^2$ have been neglected since they are very small. The bubble does not in fact contract because the surface tension forces are balanced by the force exerted by the internal excess pressure Δp. The work done in shrinkage against this pressure would be $\Delta p \times$ *area of the bubble* $\times$ *distance contracted*, i.e. $\Delta p 4\pi r^2\,\mathrm{d}r$. It follows that

$$\Delta p 4\pi r^2\,\mathrm{d}r = 8\pi r\,\mathrm{d}r\gamma$$

or

$$\Delta p = 2\gamma/r. \qquad \blacktriangleleft\textbf{(3.1)}$$

A spherical bubble has only the one radius of curvature. For non-spherical surfaces, which can be described by two radii of curvature r_1 and r_2, (3.1) takes the form

$$\Delta p = \gamma\left(\frac{1}{r_1}+\frac{1}{r_2}\right), \qquad \blacktriangleleft\textbf{(3.2)}$$

which is known as the Laplace equation. It is easy to see from (3.1) that for Δp to be positive, r must be positive and therefore the pressure is always greater on the concave side of the interface irrespective of whether or not this is a condensed phase.

A system of particular interest is that where the curved surface is the meniscus of a liquid in a cylindrical capillary tube. If the angle of contact, θ, which the liquid makes with the wall is less than 90° the liquid will rise up the capillary tube when the latter is partly immersed (fig. 3.1). If the capillary is of sufficiently small bore the meniscus will be spherical in profile. The radius of curvature of the meniscus, r, is related to the radius of the capillary tube, r_t, by

$$r_t = r\cos\theta$$

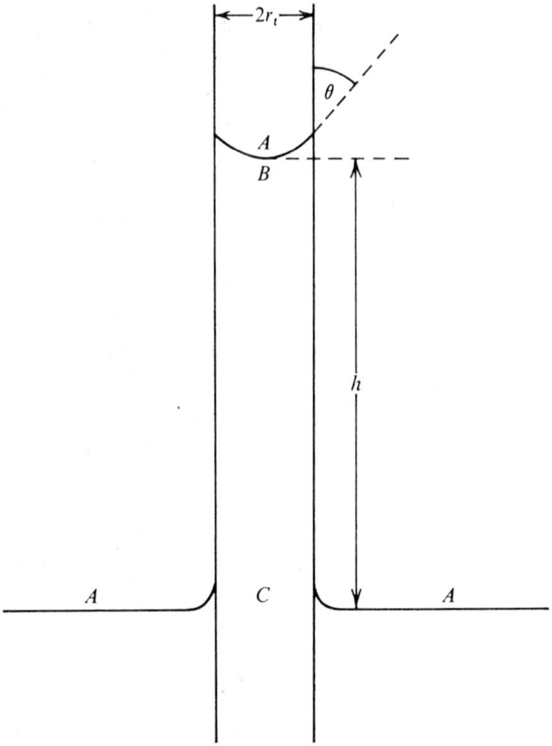

Fig. 3.1. The rise of a liquid in a capillary tube.

and therefore for this system (3.1) may be written

$$\Delta p = \frac{2\gamma \cos \theta}{r_t}. \tag{3.3}$$

With reference to fig. 3.1, the pressure at points A will be atmospheric and the length h of the column of liquid will be such that the pressure at C will also be atmospheric. The pressure at B, just below the meniscus will, according to (3.3), be less than atmospheric by an amount $2\gamma \cos \theta/r_t$. The pressure difference between A and B is equal to that between B and C. Thus

$$\frac{2\gamma \cos \theta}{r_t} = h\Delta\rho g, \tag{3.4}$$

where $\Delta\rho$ is the difference between the densities of the vapour and the liquid and $h\Delta\rho g$ is the hydrostatic pressure at C due to the column of

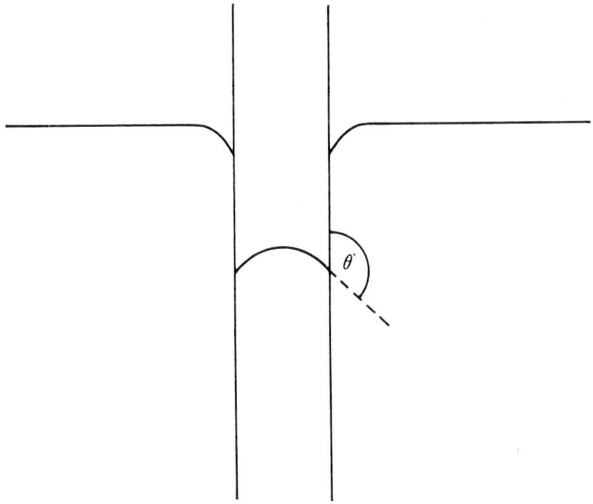

Fig. 3.2. The depression of a liquid in a capillary tube.

liquid of height h above it. The effect of the liquid above the lowest level of the meniscus has been neglected, but is mentioned in §3.3.

Suppose that a liquid makes an angle $\theta > 90^\circ$ with the wall of the capillary. The meniscus is now convex towards the vapour phase and the pressure on the liquid side of the meniscus is the greater. This leads to the depression of the liquid in the capillary, as shown in fig. 3.2. This situation is well illustrated by mercury in glass where θ is approximately 140°. Mercury is employed in the evaluation of the porosity of solid adsorbents and the use of the Laplace equation in this technique (mercury porosimetry) is discussed in §5.20.

An important aspect of the treatment of curved surfaces is that which describes the variation of the vapour pressure with curvature. This effect, as stated earlier, is formulated in the Kelvin equation. If a bulk volatile liquid with a plane surface is atomised to form many small droplets, a large area of liquid–vapour interface is created. The process therefore requires that work be done on the system and, in consequence, the chemical potential of the material in the droplets is greater than that in the non-dispersed liquid. The vapour pressure over a convex liquid surface is thus greater than that over a plane surface.

The relationship between the vapour pressure over a curved surface and the curvature of the surface ($1/r$) may be derived in the following way (Lewis & Randall, 1961).

The expression (1.20) for the differential of the Gibbs free energy

$$\mathrm{d}\mathcal{G} = -S\mathrm{d}T + V\mathrm{d}p + \gamma\mathrm{d}\mathcal{A} + \sum_i \mu_i \mathrm{d}n_i \tag{1.20}$$

was given for a plane surface, but it is applicable to a curved surface so long as γ remains unchanged by the curvature. The chemical potential in (1.20) is defined by

$$\mu_i = \left(\frac{\partial \mathcal{G}}{\partial n_i}\right)_{T,p,n_j,\mathcal{A}}. \tag{1.30}$$

This definition is convenient for a process in which n_i can be varied at constant $\mathcal{A}$, e.g. the transfer of material across a plane interface. For a small droplet, however, the transfer of material necessarily results in a change in $\mathcal{A}$. Thus, when $\mathrm{d}n_i$ moles of component i are added to a droplet the volume change, $\mathrm{d}V$, of the droplet is

$$\mathrm{d}V = \sum_i V_i \mathrm{d}n_i,$$

where V_i is the partial molar volume of component i in the liquid. The volume of the droplet of radius r is $\frac{4}{3}\pi r^3$ and so $\mathrm{d}V$ is $4\pi r^2 \mathrm{d}r$. The change $\mathrm{d}\mathcal{A}$ in surface area of the droplet is $8\pi r\,\mathrm{d}r$. It follows that

$$\mathrm{d}\mathcal{A} = \frac{2\mathrm{d}V}{r} = \sum_i \frac{2V_i}{r}\mathrm{d}n_i. \tag{3.5}$$

Combining (1.20) and (3.5),

$$\mathrm{d}\mathcal{G} = -S\mathrm{d}T + V\mathrm{d}p + \gamma\sum_i \frac{2V_i}{r}\mathrm{d}n_i + \sum_i \mu_i \mathrm{d}n_i \tag{3.6}$$

or

$$\mathrm{d}\mathcal{G} = -S\mathrm{d}T + V\mathrm{d}p + \sum_i \left(\frac{2V_i\gamma}{r} + \mu_i\right)\mathrm{d}n_i. \tag{3.7}$$

The chemical potential μ_i' of component i in the droplet is therefore

$$\mu_i' = \left(\frac{\partial \mathcal{G}}{\partial n_i}\right)_{T,p,n_j}$$

or

$$\mu_i' = \frac{2V_i\gamma}{r} + \mu_i.$$

Thus

$$\mu_i' - \mu_i = \frac{2V_i\gamma}{r}.$$

But

$$\mu_i = \mu_i^0 + RT\ln p_i^0,$$

where p_i^0 is the vapour pressure of i over a plane surface, and

$$\mu_i' = \mu_i^0 + RT\ln p_i,$$

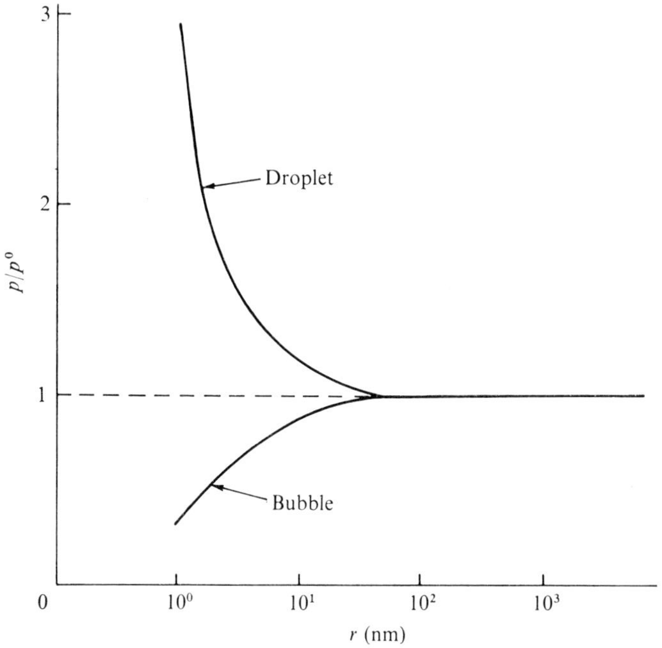

Fig. 3.3. Effect of curvature on the vapour pressure of water. It has been assumed for these calculations that the surface tension is independent of curvature.

where p_i is the vapour pressure over the curved surface. Therefore

$$\ln \frac{p_i}{p_i^0} = \frac{2V_i\gamma}{rRT}.$$

For a one component system

$$\ln \frac{p}{p^0} = \frac{2V_m\gamma}{rRT}, \tag{3.8}$$

where V_m is the molar volume of the liquid. Equation (3.8) is a form of the Kelvin equation. For a bubble of vapour in a liquid the radius of curvature takes the opposite sign and (3.8) becomes

$$\ln \frac{p}{p^0} = -\frac{2V_m\gamma}{rRT}. \tag{3.9}$$

Fig. 3.3 shows the effect of curvature on the vapour pressure of water, both for convex menisci (droplets) and concave menisci (bubbles). It can be seen that curvature has little effect on the vapour pressure

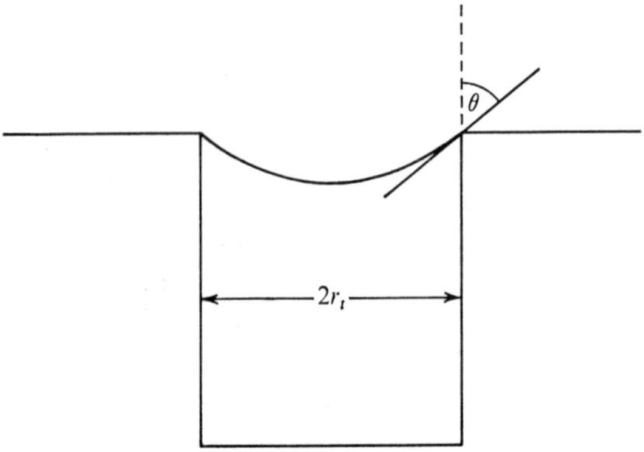

Fig. 3.4. Liquid in a cylindrical capillary.

until the radius is of the order of 10 nm (100 Å) but it must be remembered that the Kelvin equation is a thermodynamic relationship which may be invalid for systems of molecular dimensions.

For non-spherical surfaces, the term $2/r$ which appears in (3.8) and (3.9) must be replaced by $(1/r_1+1/r_2)$ where r_1 and r_2 are the two radii of curvature of the surface.

It is instructive to consider (3.9) in conjunction with (3.1) when applied to the formation of small bubbles of vapour in a liquid. Suppose the liquid is under a pressure p'. At the boiling point of the liquid the vapour pressure (p^0) over the plane surface of the liquid equals p'. If a bubble were to be formed, the pressure p'' inside the bubble would have to be, from (3.1), greater than p'. This situation, however, is not permitted by (3.9) because, as the right-hand side is negative, $p''(=p)$ must be less than $p'(=p^0)$. Bubbles cannot therefore exist at the boiling point. It can be shown that for the formation of bubbles it is necessary to superheat the liquid.

The treatment of the vapour pressure over curved surfaces also finds application in studies of the supersaturation of vapours and solutions (see Adamson, 1967*a*; Defay, Prigogine, Bellemans & Everett, 1966*a*). The Kelvin equation has, in addition, been extensively used in studies of the adsorption of gases and vapours on to porous solids. Fig. 3.4 represents liquid with a spherically curved interface in a cylindrical capillary of radius r_t. Equation (3.9) for this situation becomes

$$\ln\frac{p}{p^0} = -\frac{2V_m\gamma\cos\theta}{r_t RT}. \tag{3.10}$$

This expression indicates the pressure p at which a vapour will condense in the capillary. If, as depicted, $\theta < 90°$, so that the meniscus is concave, $p < p^0$. If, on the other hand, $\theta > 90°$, $p > p^0$, i.e. vapour will condense on the plane surface first (at p^0) and a pressure greater than p^0 is required to force liquid into the capillary. A fuller discussion of these applications is given in §§ 5.18–20.

So far it has been assumed that the surface tension is independent of curvature, and, indeed, for most systems of interest ($r > 100$ nm) this is likely to be so. However, when the radius of curvature is comparable to molecular dimensions, calculations suggest that the surface tension becomes dependent on this quantity. Thus it has been calculated that for a water droplet of radius 1 nm the surface tension is only 0.755 of that for a plane surface (Defay, Prigogine, Bellemans & Everett, 1966*b*).

3.3. The measurement of surface and interfacial tension. Three methods only will be discussed. A much wider treatment of this subject has been given by Adamson (1967*b*).

The *capillary rise method* is generally considered to be the most accurate available provided that certain precautions are taken in its use. Its scope, however, is rather limited, and the method is most suited to the measurement of the surface tension of pure liquids. For dilute aqueous solutions of surface active materials other methods are preferable.

The simple theory of capillary rise has been discussed in § 3.2 where (3.4) shows the relation between the height h, which a liquid ascends a capillary, and the surface tension γ of the liquid. Rearranging (3.4),

$$\gamma = \frac{r_t h \Delta\rho g}{2\cos\theta}. \tag{3.11}$$

For practical reasons, if the method is to be accurate, θ must be zero; i.e. the liquid must completely wet the material of the capillary. Contact angles which are not zero are difficult to determine and to reproduce. For narrow capillaries, say 0.1 mm radius, where $\theta = 0$, the liquid meniscus may be assumed to be hemispherical and (3.11) becomes simply

$$\gamma = \frac{r_t h \Delta\rho g}{2}. \qquad \blacktriangleleft\textbf{(3.12)}$$

Fig. 3.5 shows a simple form of a capillary rise apparatus. The height of the liquid in the capillary (A) and the plane liquid in the wide tube

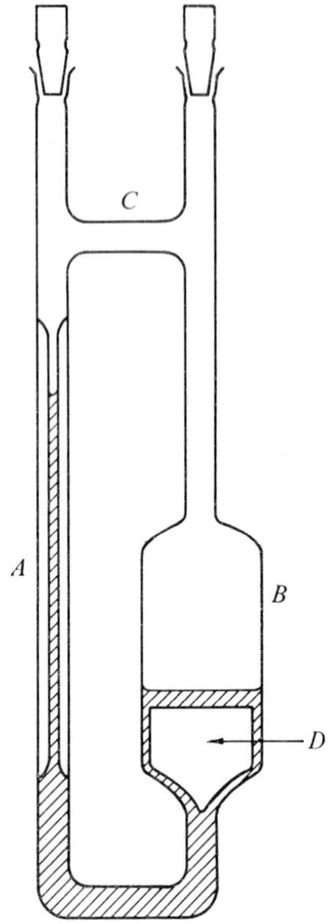

Fig. 3.5. Simple capillary rise apparatus.

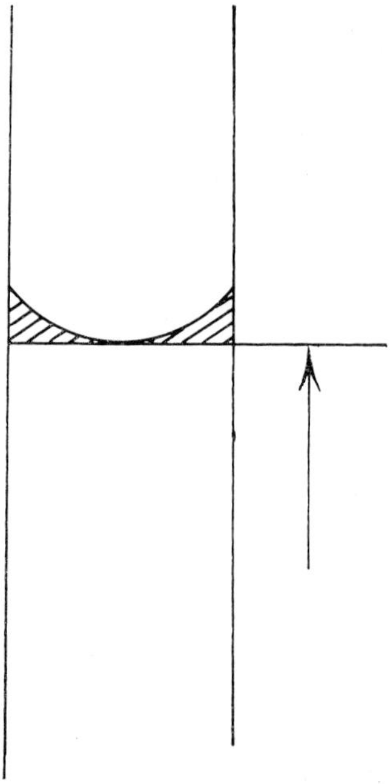

Fig. 3.6. Liquid above the level of the bottom of the meniscus.

(B) is measured by means of a cathetometer. The radius of B is made sufficiently large for the surface of the liquid at the centre to be planar. A float D may be included in B so as to reduce the volume of liquid required for the experiment. The cross tube C ensures that the pressure in both arms of the apparatus is the same.

The height h of liquid in A is measured to the bottom of the meniscus, as shown in fig. 3.6. This means that the hydrostatic pressure of the liquid above this level is not taken into account. It can be shown, however, that for narrow capillaries it is a sufficiently good correction simply to add $r_t/3$ on to h. This meniscus correction is normally very small.

The *drop-volume method* involves the determination of the volume, or weight, of a drop of liquid surrounded by vapour or by a second liquid, as it becomes detached from a tip of known radius. Fig. 3.7 shows a partly formed drop of liquid, whose surface tension is to be determined, hanging from a vertically mounted tip. To a very crude approximation one may write, for the weight W of a drop which just becomes detached,

$$W = 2\pi r\gamma, \tag{3.13}$$

where r is the radius of the tip. There are several respects in which (3.13) is inadequate, perhaps the most obvious being that in practice the whole of the drop never falls from the tip. The theory of the detachment of drops is not complete, but an empirical calibration of the drop-volume method has been given by Harkins & Brown (1919). These workers determined the weights of drops of water and of benzene which fell from tips of different sizes and, in addition, determined the surface tensions of these liquids by means of the capillary rise method. They were thus able to compute correction factors ϕ, which corresponded to various drop volumes V and tip radii, for inclusion in (3.13), i.e.

$$W = 2\pi r\gamma\phi,$$

where $$\phi = f(r/V^{\frac{1}{3}}).$$

Equation (3.13) may now be written

$$\gamma = \frac{V\Delta\rho g}{2\pi r\phi}, \tag{◀3.14}$$

where $\Delta\rho$ is the difference in density between the drop and its surroundings.

Although the drop-volume method is calibrated for the liquid–vapour interface it is commonly used (and indeed is extremely convenient) for measurements of interfacial tension. Where interfacial tensions of oil–water systems have been determined by the drop-volume method and by the Wilhelmy plate method (q.v.) the agreement has been satisfactory.

A simple apparatus for the accurate determination of drop volume is illustrated in fig. 3.8. The liquid is contained in a precision made syringe A (capacity *ca.* 500 mm^3) the plunger of which is operated by a micrometer head B. The liquid is forced from the tip C (radius *ca.* 2–3 mm) which is usually made of stainless steel or glass and which is accurately circular and of known radius. The drop D is surrounded by vapour or by a second liquid and the apparatus is suitably thermostatted.

This method gives very reproducible results for surface and interfacial tensions in systems of pure components, liquid mixtures and dilute

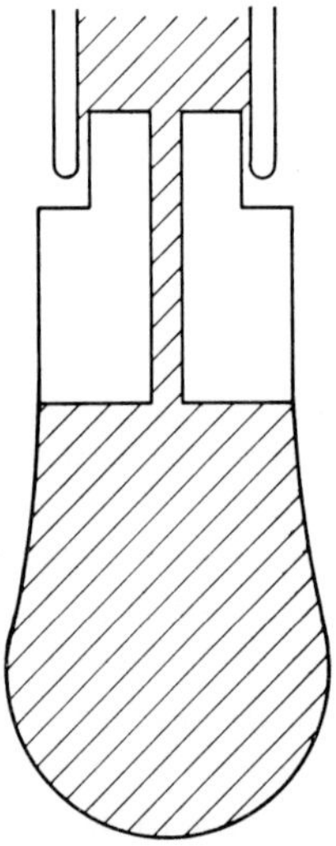

Fig. 3.7. Partially formed liquid drop hanging from a vertically mounted tip.

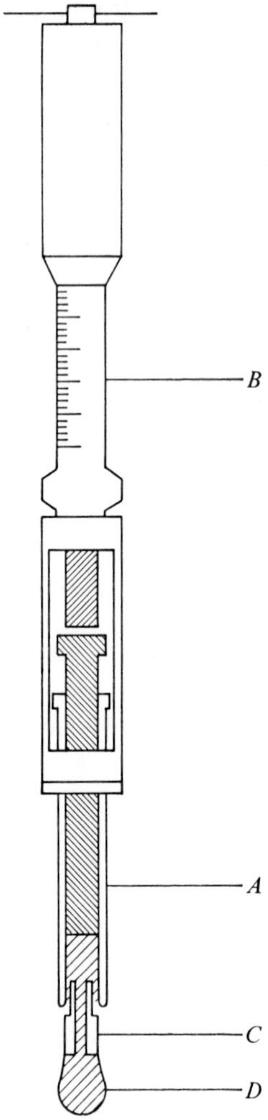

Fig. 3.8. Drop-volume apparatus.

solutions of surface active agents. For this reason it has become very popular even though it is not an absolute method.

The *Wilhelmy plate method* is illustrated in fig. 3.9. A thin plate of glass, mica or platinum is suspended from one arm of a balance and is partially immersed in the liquid whose surface tension is to be deter-

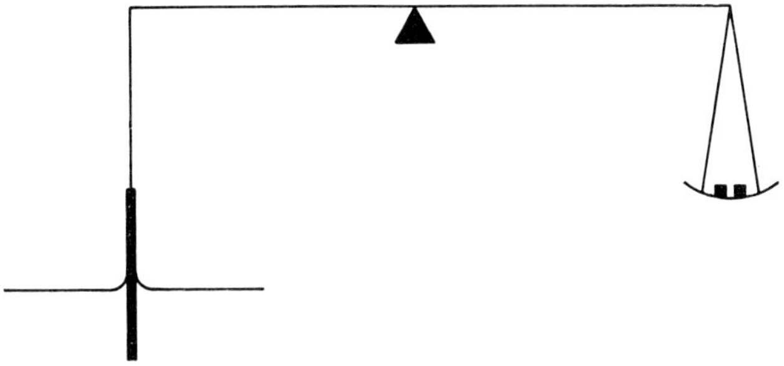

Fig. 3.9. Schematic diagram of the Wilhelmy plate method.

mined. In one modification of the method the liquid is lowered until the plate becomes detached from the surface and the maximum pull W_t on the balance is noted. The difference $(W_t - W)$ between this pull and the weight W of the plate in air is the weight of the liquid in the meniscus. If it is arranged, by suitable treatment of the plate surface, that the contact angle of the liquid with the plate is zero, then

$$W_t - W = P\gamma,$$

where P is the perimeter of the plate. P can normally be set equal to twice the length of the plate. When the method is used for measurements on insoluble monolayers it is usually not desirable to detach the plate. Instead, the force necessary to keep the plate at a constant depth of immersion as the surface tension is altered, is determined. In this modification also a zero contact angle is necessary, and a correction for the buoyancy of the plate may be necessary. The chief difficulty of the Wilhelmy plate method lies in the establishment of zero contact angle, but subject to this the method is absolute. It is of particular value in the study of insoluble monolayers at oil–water interfaces (Brooks & Pethica, 1964).

3.4. Surfaces of pure liquids. Some surface and interfacial tensions between pure liquids and their vapours and between immiscible or partially miscible liquids are given in table 3.1. Against their vapours and at room temperature pure liquids usually have tensions in the range 10–80 mN m^{-1}, water being in the upper and organic liquids in the lower parts of this range. Interfacial tensions between hydrocarbons and water are intermediate between the surface tensions of the pure

TABLE 3.1. *Surface and interfacial tensions of some liquids* at* 293 K

	Liquid–vapour	Water–liquid	$-d\gamma/dT$ (liquid–vapour)	$-d\gamma/dT$ (water–liquid)
Water	72.75	—	0.16	—
Octane	21.69	51.68	0.095	0.09
Dodecane	25.44	52.90	0.088	0.09
Hexadecane	27.46	53.77	0.085	—
Benzene	28.88	35.00	0.13	—
Carbon tetrachloride	26.77	45.0	—	—
Mercury	476	375	—	—

* All values of γ are in mN m^{-1}.

components. The relationship between surface and interfacial tensions in such systems is discussed in §3.5. Liquid metals are often exceptional in their surface properties and may have tensions as high as several hundred mN m^{-1}.

With increasing temperature the surface tension of most liquids decreases (table 3.1). A negative temperature coefficient for the tension indicates that the surface (excess) entropy is positive. This is evident from (1.37). If the Gibbs model for the surface is used, (1.37) for a one component system and unit area of surface becomes

$$S_\sigma dT + d\gamma + \Gamma^\sigma d\mu = 0. \tag{3.15}$$

When the position of the dividing surface is chosen such that Γ^σ, the surface excess, is zero, (3.15) reduces to

$$\frac{d\gamma}{dT} = -S_\sigma, \qquad \blacktriangleleft\textbf{(3.16)}$$

where S_σ is the specific surface excess entropy, i.e. the entropy of unit area of surface liquid less the entropy of the same amount of bulk liquid. An equivalent expression may be obtained by use of the alternative (surface phase) model. The explanation at a molecular level for a positive surface excess entropy is not properly understood. It is intuitively reasonable to expect that molecules in a surface may have more freedom of movement or are more disordered than in the bulk, but it is not clear what type of model should be used for the calculation of the configurational entropy. This point is mentioned again below.

There are several empirical expressions for the surface tension of pure liquids. The best known are those of Eotvös and of Ramsay and Shields. The chief value of these expressions is that they facilitate the calculation of approximate values of the surface tension and surface entropy of a liquid from certain other physical properties such as critical temperature and molar volume. Their theoretical foundation is, on the whole, rather obscure and for further discussion the reader is referred to Guggenheim (1967).

Calculations of the surface tension, energy, entropy etc. from fundamentals have been attempted by the use of statistical thermodynamics. At present the use of these methods does not yield very accurate values of the thermodynamic parameters, even for the simplest systems. Such attempts as have been made are nevertheless of considerable importance and some description of the present situation is appropriate. There are two general approaches to the problem (Hill, 1960; Defay, Prigogine, Bellemans & Everett, 1966*c*). The first is the so-called radial distribution function method and the second is based on the cell theory of liquids proposed originally by Lennard-Jones and Devonshire. By either method it is possible, in principle, to calculate the surface excess free energy and hence to obtain the surface tension from the expression

$$\gamma = \left(\frac{\partial A^{\sigma}}{\partial \mathcal{A}}\right)_{T,V,N}. \tag{3.17}$$

Owing to the operation of intermolecular forces, the distribution or density of molecules surrounding a given molecule in a liquid does not average out to the bulk value until at least the second or third nearest neighbours are reached. For spherical molecules where, for example, the pair-wise potential energy is given by, say, the Lennard-Jones 6–12 formula, the greatest density of neighbours will be found at a distance corresponding to the minimum in this potential energy curve. The density of molecules at various distances from a central molecule is directly proportional to a quantity known as the radial distribution function. Some knowledge of this function may be obtained from X-ray diffraction data. If, together with this information, a potential function (such as that of Lennard-Jones) is assumed, the surface tension may be calculated. In this way Kirkwood and Buff calculated the surface tension of liquid argon at 90 K and obtained a value of 14.9 mN m^{-1} as compared with the experimental value of 11.9 mN m^{-1}.

The application of the cell model of liquids to the calculation of surface tension involves the derivation and evaluation of the partition function of the molecules in the surface and bulk of the liquid. The liquid is considered to be divided up into cells such that there is one molecule per cell although, as will be mentioned below, there is an important modification of this approach in which the existence of empty cells (holes) is invoked. The cell is formed by a cage of nearest neighbours and within this cage the molecule oscillates in the field generated by its neighbours. This oscillation is assumed independent of the motion of the other molecules. The partition function is expressed as two terms, one of which is for the stationary molecules in their potential minima at the centre of their cells and the second of which is for their kinetic and potential energy at other points in the cell. The difference in the numbers of nearest neighbours for bulk and surface molecules leads to different partition functions for the bulk and surface regions. From the combined partition function Z the free energy of the system can be obtained from the expression

$$A = -kT \ln Z \tag{3.18}$$

and the surface tension is easily obtained from A if the surface area per molecule is known. The calculation of the partition function turns largely on the way in which the potential energy is assumed to vary with position in the cell. For a square-well potential Prigogine and Saraga calculated the surface tension of liquid argon at 85 K to be 9 mN m^{-1}. This may be compared with the observed value of 13.2 mN m^{-1}. A marked improvement in the agreement was obtained, however, by repeating the calculations with the assumption that the surface layer of molecules was incomplete. The existence of holes in the surface layer leads, as one might expect, to greater freedom of movement for the molecules adjacent to the holes and to a larger number of possible arrangements of the molecules in the layer. Both of these effects tend to enhance the surface entropy, energy and, on balance, also the surface tension.

More detailed discussions of both the radial distribution function and cell model calculations are given by Hill (1960) and Defay, Prigogine, Bellemans & Everett (1966*c*).

The sharpness of the surface is a matter of some importance for the calculation of the surface tension. Whether the transition from liquid to vapour occurs over one or over several molecular diameters depends on the effective range of the intermolecular forces. These forces are inversely proportional to the seventh power of the distance and it is therefore to be expected that their range is no more than a few ångström units. In the calculations of the surface tension described above only nearest neighbour interactions are considered, and reasonable values are obtained. The transition from liquid to vapour probably occurs, therefore, over no more than one or perhaps two molecular diameters. This picture is supported by optical measurements. According to Fresnel's law, light incident at the Brewster angle is plane polarised on reflection from an infinitely sharp surface. Such measurements as have been carried out for liquids in contact with gases show that the boundary is very sharp. It is nevertheless important to remember that the surface of a liquid is by no means a static structure and that the exchange of molecules between the surface and the adjacent bulk phases is a rapid process. The rest time of a molecule in the surface of a liquid is usually in the region of 10^{-7} s, as may be deduced from the kinetic theory of gases.

When molecules are not spherically symmetrical it is to be expected that they will have a preferred orientation at an interface. For example, for the interface between a liquid n-alkanol and water it seems intuitively likely that the alkanol molecules will be oriented in the interface such that their hydroxyl groups will be in contact with the water, and their hydrocarbon chains extended into the alkanol. This conclusion may be reached in a more convincing manner from the consideration of standard free energies of adsorption in three component systems. When n-alkanols adsorb from water to a hydrocarbon–water interface the standard free energies of adsorption become more negative by approximately 3.2 kJ mol^{-1} for each additional methylene group in the molecule. The quantity -3.2 kJ mol^{-1} is very close to the free energy of transfer per methylene group of aliphatic long chain molecules from water to liquid hydrocarbon, and hence it is concluded that on adsorption, the hydrocarbon chain of the alkanol must pass from the water into the hydrocarbon phase. If, on the other hand, the adsorption of alkanols or other polar aliphatic molecules from hydrocarbon to hydrocarbon–water interfaces is examined, it is found that the standard free energy of adsorption is largely independent of the length or structure of the hydrocarbon chain of the molecule, but does depend strongly on the nature of the polar group. From this evidence it is concluded that on adsorption from the hydrocarbon side of the interface only the polar group of the molecule transfers to the aqueous phase. As, at equilibrium, the orientation of an alkanol molecule in the interface must be independent of the phase from which it is adsorbed the combined evidence confirms the intuitive conclusion that the molecules have their polar groups in the water and their hydrocarbon chains in the oil phase. Although this conclusion would perhaps be less valid for a pure alkanol–water interface, especially if the alkanol was of relatively short chain length, the general principles governing the orientation of this type of asymmetric molecule are well illustrated by the example.

The orientation of water at an interface is more difficult to establish. The approach which has on several occasions been attempted rests on the estimation of the surface potential which should arise if the water dipoles have a predominant orientation at the interface. However, as pointed out in §2.2, this potential cannot be measured. It can be inferred from extra-thermodynamic considerations, but is subject to some uncertainty.

3.5. Spreading and adhesion in liquid–liquid systems. The study of the spreading of one liquid on a second liquid, and of the adhesion of two liquids, is of considerable interest in its own right and may also throw light on problems associated with the wetting of solids by liquids. Many of the arguments are equally applicable to either type of system but the heterogeneity and roughness of solid surfaces introduces further complications. The similarities between liquid–liquid and liquid–solid systems have been pointed out frequently (e.g. Johnson & Dettre, 1966) and relationships presented below will be referred to in §6.2, which is concerned with the solid–liquid interface.

If a small amount of a liquid α is placed on the plane surface of an immiscible (or only slightly miscible) liquid β, one might ask whether α will spread over the surface of β and, if so, in what kind of film and, in any event, how strongly α and β adhere. To discuss such questions two quantities are defined, namely *the work of cohesion*, W_C, of a liquid and the reversible work of separation per unit area (often called the *work of adhesion*) W_A, of two liquids. Suppose that a column of liquid, with unit cross sectional area is split in such a way as to give two units of area of equilibrium surface. This requires work W_C where

$$W_C = 2\gamma. \quad ◄(3.19)$$

Now suppose that a column of liquid α with unit cross section is resting on a similar column of liquid β, giving unit area of $\alpha\beta$ interface with an interfacial tension $\gamma^{\alpha\beta}$. The reversible work required to part the two columns, so as to form unit surface areas of α and β, is given by the Dupré equation

$$W_A^{\alpha\beta} = \gamma^{\alpha}+\gamma^{\beta}-\gamma^{\alpha\beta}. \quad ◄(3.20)$$

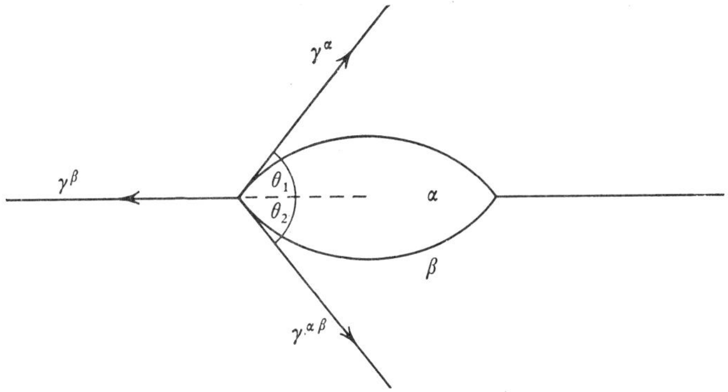

Fig. 3.10. A non-spreading liquid α on liquid β.

When a drop of α is placed on the surface of β, one of several things may happen. Phase α may remain as a lens as shown schematically in fig. 3.10. Alternatively, it may spread either as a monolayer (with or without a residual lens) or as a relatively thick film which, if γ^α and $\gamma^{\alpha\beta}$ retain their bulk values, is often called a duplex film.

If fig. 3.10 is to represent an equilibrium situation, α and β must be mutually saturated and hence the surface and interfacial tensions will not necessarily be those of the pure components. At hydrostatic equilibrium it is evident that

$$\gamma^\beta = \gamma^\alpha \cos\theta_1 + \gamma^{\alpha\beta}\cos\theta_2. \tag{3.21}$$

It is very useful to consider the hypothetical situation in which the surface of β remains planar beneath the drop of α. The examination of such a system leads to relationships which may be applied to the wetting of solids by liquids. By reference to fig. 3.11 the condition for hydrostatic equilibrium is

$$\gamma^\beta = \gamma^{\alpha\beta} + \gamma^\alpha \cos\theta. \qquad \blacktriangleleft\textbf{(3.22)}$$

This equation is usually called Young's equation. The parameter θ, termed the *contact angle*, is defined by (3.22) and is obviously not experimentally observable in liquid–liquid systems. As a physical concept θ presents little difficulty so long as it is finite. However, the significance of $\theta = 0$ is somewhat obscure as it implies complete spreading and, when this occurs, there are two clearly distinguishable possibilities. Firstly, the spreading liquid may be contained by the walls of the vessel so that it forms a thick layer. In this instance the three

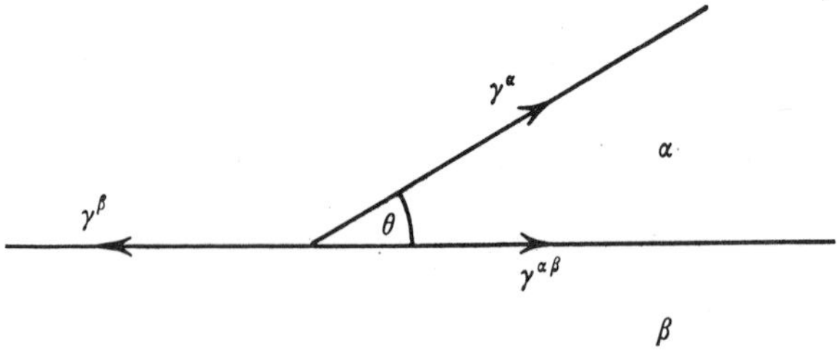

Fig. 3.11. See text.

phase boundary (i.e. α–β–vapour) vanishes and the contact angle is non-existent. Secondly, the spreading liquid, if not restrained, may expand to a film of molecular thickness. Provided that this film does not completely cover the underlying liquid it may be said that there is a boundary between α, β and the vapour. However, the use of θ, which is a macroscopic parameter, to describe this boundary is clearly inappropriate.

In order to discuss the phenomenon of spreading of thick films in numerical terms, the concept of the *spreading tension* (frequently called the *spreading coefficient*), σ, was introduced by Harkins. In the case where liquids α and β are mutually saturated, $\sigma^{\alpha\beta}$, for the spreading of α on β, is defined as

$$\sigma^{\alpha\beta} = \gamma_e^\beta - \gamma_e^\alpha - \gamma_e^{\alpha\beta}, \tag{3.23}$$

where it is convenient to denote mutual saturation by subscript 'e'. From (3.19), (3.20) and (3.23),

$$\sigma^{\alpha\beta} = (W_A^{\alpha\beta})_e - (W_C^\alpha)_e, \qquad \blacktriangleleft\textbf{(3.24)}$$

where $(W_C^\alpha)_e$ represents the work of cohesion of α saturated with β. If the adhesion between α and β is greater than the cohesion of α, spreading will take place, and from (3.24) $\sigma^{\alpha\beta} > 0$. Conversely, if $(W_C^\alpha)_e > (W_A^{\alpha\beta})_e$ spreading will not occur and $\sigma^{\alpha\beta} < 0$.

If a drop of pure α is placed on pure β it will normally take an appreciable time for mutual saturation to take place and for this reason it is convenient to define an *initial spreading tension*, $\sigma_i^{\alpha\beta}$ such that

$$\sigma_i^{\alpha\beta} = \gamma^\beta - \gamma^\alpha - \gamma^{\alpha\beta}, \tag{3.25}$$

where γ^α and γ^β are the surface tensions of pure α and β, and $\gamma^{\alpha\beta}$ the original interfacial tension. It is sometimes observed that a liquid initially spreads (i.e. $\sigma_i^{\alpha\beta} > 0$) but that after a time a retraction occurs

($\sigma^{\alpha\beta} < 0$). A well-known example of this phenomenon is that of benzene on water where the mutual solubility is appreciable. For this system $\sigma_i^{\alpha\beta} = +8.9$ mN m^{-1} whereas $\sigma^{\alpha\beta} = -1.4$ mN m^{-1} and the final (equilibrium) system consists of lenses of benzene in equilibrium with an adsorbed monolayer. It is worth noting that many solids are able to spread on the surface of water as monolayers (§3.7).

A recent concept, originally developed from experiments performed on solid–liquid systems, is that of the *critical surface tension*, γ_C. It is useful in the characterisation of both liquid and solid surfaces and is explained here by reference to liquid alkane (α)–water (β) systems (Shafrin & Zisman, 1967). At 20 °C all the n-alkanes below nonane spread spontaneously on water ($\sigma_i^{\alpha\beta} > 0$) but the n-alkanes nonane and upwards are non-spreading liquids. Thus the value of γ_C for the spreading of alkanes on water lies between the surface tensions of octane and nonane, which are 21.8 and 22.9 mN m^{-1} respectively at 20 °C. The actual value of γ_C can be obtained from a plot, which is rectilinear, of $\sigma_i^{\alpha\beta}$ versus γ^α. For $\sigma_i^{\alpha\beta} = 0$, $\gamma^\alpha = \gamma_C$, so that γ_C can be obtained by interpolation. Shafrin and Zisman, by the use of this method, obtained a value for γ_C of 21.9 mN m^{-1}. Since $\sigma_i^{\alpha\beta}$ (rather than $\sigma^{\alpha\beta}$) was used this value of γ_C refers to non-equilibrium systems. Johnson & Dettre (1966), however, obtained a value of 19.1 mN m^{-1} for alkane–water systems at saturation equilibrium at 24.5 °C.

The same value for γ_C as that obtained using n-alkanes is also found for the spreading of branched alkanes and cyclic saturated hydrocarbons (naphthenics). It thus appears that γ_C is a property of the water alone, provided that the other liquids are saturated hydrocarbons, and is therefore referred to as the critical surface tension of water. When unsaturated hydrocarbons are used, however, higher values of γ_C are obtained. Shafrin and Zisman suggest that while for saturated hydrocarbons γ_C is determined only by dispersion forces, for unsaturated hydrocarbons forces other than dispersion forces contribute to the liquid–liquid adhesion. Thus, although γ_C is referred to as a property of the substrate, it is also dependent in some measure on the second liquid.

Attempts have been made to interpret wetting and adhesion phenomena in terms of molecular forces (Girifalco & Good, 1957; Fowkes, 1963, 1964) but the treatments are semi-empirical. An assumption in the work of Fowkes is that the surface tension of, say, water can be expressed as the sum of contributions originating from dispersion forces, γ^d, and from other forces including those generated by hydrogen bonding, γ^h. It can be shown that on Fowkes' model, γ_C for water against saturated hydrocarbons, is simply γ^d for water. However, the state of the theory of molecular interactions at interfaces is still largely undeveloped and the critical surface tension can only be regarded as an empirically derived quantity.

3.6. Surfaces of binary liquid mixtures. The theory of the surfaces of binary liquid mixtures is relatively well developed. In such systems it is usual for there to be marked similarities between the two types of molecule and it is uncommon for either to be strongly adsorbed. The extreme example of this type of mixture is the perfect solution. In this section will be developed equations which relate the surface tensions of mixtures to their composition and to the surface tension of the pure components, both for perfect solutions and for athermal (zero heat of mixing) mixtures of molecules of different size.

In practice, adsorption from liquid mixtures is not of such wide interest as adsorption from dilute solutions. The treatment of dilute solutions (e.g. of strongly surface active molecules) is in general, however, rather complicated and poorly understood, and before encountering this topic in §3.8 it is useful to be familiar with the results for simple liquid mixtures.

In the two treatments of mixtures to be outlined it is assumed that the surface is a monolayer, i.e. that only this layer differs in composition from the underlying bulk solution. This model is probably an over-simplification of the reality but it serves to make the problem easier. The model allows the composition of the surface to be specified simply in terms of the molar surface areas and numbers of moles of each species. The surface area $\mathcal{A}$ is given by

$$\mathcal{A} = \sum_i n_i^m a_i. \tag{3.26}$$

n_i^m is the number of moles of i in the monolayer; it replaces n_i^σ in the general thermodynamic expressions for surfaces given in §1.7. The parameter a_i is the partial molar surface area of i defined in §1.7 as

$$a_i = \left(\frac{\partial \mathcal{A}}{\partial n_i^m}\right)_{T, V, \gamma, n_j^m}. \tag{3.27}$$

To a near approximation, a_i is the area occupied by one mole of component i spread as a compact monolayer (i.e. two-dimensional liquid). For a binary system, (3.26) may be written

$$n_1^m a_1 + n_2^m a_2 = \mathcal{A}. \tag{3.28}$$

Usually, a_1 and a_2 are supposed to be independent of the composition of the monolayer and are put equal to the molar surface areas.

The general procedure for the derivation of a surface tension equation is to obtain suitable expressions for the bulk and surface chemical potentials of a species. At adsorption equilibrium these two chemical potentials may be equated to give the required expression. The first type of system to be considered is the *perfect solution* (Defay, Prigogine, Bellemans & Everett, 1966*d*).

The chemical potential, μ_i^l of component i in an ideal solution is written

$$\mu_i^l = \mu_i^{\ominus, l} + RT \ln x_i^l, \tag{3.29}$$

where x_i^l is the bulk mole fraction of i and $\mu_i^{\ominus, l}$ is a standard chemical potential. Furthermore, it will be recalled from §1.7 that the surface chemical potential μ_i^σ is given by

$$\mu_i^\sigma = \zeta_i - \gamma a_i \tag{1.33}$$

and that at equilibrium

$$\mu_i^\sigma = \mu_i^l. \tag{1.29}$$

In the discussion of monolayers which follows, the superscript σ is replaced by superscript m. For an ideal monolayer ζ_i may be written

$$\zeta_i = \mu_i^{\ominus, m} + RT \ln x_i^m, \tag{3.30}\dagger$$

where x_i^m is the surface mole fraction of component i in the monolayer and $\mu_i^{\ominus, m}$ is a standard surface chemical potential. For the monolayer (1.33) may now be written as

$$\mu_i^m = \mu_i^{\ominus, m} + RT \ln x_i^m - \gamma a_i. \tag{3.31}$$

At equilibrium, by the use of (1.29), (3.29) and (3.31)

$$\mu_i^{\ominus, l} - \mu_i^{\ominus, m} = RT \ln \frac{x_i^m}{x_i^l} - \gamma a_i. \tag{3.32}$$

For a perfect solution (i.e. one which is ideal at all concentrations) $\mu_i^\ominus$ becomes μ_i^0, the chemical potential of the pure component. Furthermore, for a pure component

$$\mu_i^{0, l} - \mu_i^{0, m} = -\gamma_i^0 a_i, \tag{3.33}\dagger$$

where γ_i^0 is the surface tension of pure liquid. Thus, for perfect solutions (3.32) becomes

$$\gamma = \gamma_i^0 + \frac{RT}{a_i} \ln \frac{x_i^m}{x_i^l}. \qquad \blacktriangleleft\,\mathbf{(3.34)}$$

† See appendix.

TABLE 3.2. *Observed and calculated surface tensions of binary liquid mixtures at* $x_1^l = x_2^l = 0.5$

System	T (K)	a/N_A (nm^2)	γ_1^0 (mN m^{-1})	γ_2^0 (mN m^{-1})	γ (observed) (mN m^{-1})	γ (calculated from (3.37)) (mN m^{-1})
Benzene+ +m-xylene	291.2	0.377	28.40	28.40	28.40	28.40
Methanol+ ethanol	303	0.2284	21.06	20.76	20.91	20.91
H_2O+D_2O	298	0.112	72.06	70.91	71.48	71.48
m-Xylene+ o-xylene	293	0.413	28.90	30.17	29.20	29.52
m-Xylene+ p-xylene	293	0.417	28.90	28.37	28.50	28.63
Chlorobenzene+ bromobenzene	293	0.370	33.11	36.60	34.65	34.72

Or, more specifically,

$$\left.\begin{aligned}\gamma &= \gamma_1^0+\frac{RT}{a_1}\ln\frac{x_1^m}{x_1^l}\\ &= \gamma_2^0+\frac{RT}{a_2}\ln\frac{x_2^m}{x_2^l}.\end{aligned}\right\} \tag{3.35}$$

If $a_1 = a_2 = a$, which should be approximately true for perfect solutions, (3.35) yields

$$\left.\begin{aligned}x_1^m &= x_1^l \exp\,[a(\gamma-\gamma_1^0)/RT],\\ x_2^m &= x_2^l \exp\,[a(\gamma-\gamma_2^0)/RT].\end{aligned}\right\} \tag{3.36}$$

If therefore a, γ_1^0 and γ_2^0 are known, the composition of the surface may be calculated from a knowledge of the surface tension of the mixture. The component with the lower surface tension will obviously be preferentially adsorbed.

For the special case in which $a_1 = a_2 = a$, (3.35) may be written as

$$\exp\left(-\frac{\gamma a}{RT}\right) = x_1^l \exp\left(-\frac{\gamma_1^0 a}{RT}\right)+x_2^l \exp\left(-\frac{\gamma_2^0 a}{RT}\right). \tag{3.37}$$

This expression, which was first presented by Guggenheim (1945), is more convenient than (3.35) for the calculation of γ from x_1^l, x_2^l, γ_1^0 and γ_2^0. Belton & Evans (1945) have compared experimental results with the predictions of the theory and have found reasonable agreement for several systems. In table 3.2 some observed and calculated surface tensions are listed for $x_1^l = x_2^l = 0.5$. Also given in the table are γ_1^0 and γ_2^0 together with values of a calculated from bulk liquid densities. Analogues of (3.35) will be referred to later in connection with the adsorption from liquid mixtures at the solid–liquid interface.

Owing to the assumption that the solutions were perfect, the above treatment is appropriate only for molecules of similar size. For mixtures of molecules of very different sizes the treatment is necessarily more complex, although a relatively simple statistical approach has been given for certain types of system by Prigogine & Maréchal (1952). These authors have derived equations for binary liquid mixtures of molecules of different sizes, which relate the surface tensions of the mixtures to those of the two pure components, the composition of the liquid and the ratio of the sizes of the molecular species.

The system envisaged consists of a solution containing N_1 monomer molecules each occupying one site, and N_2 chain-like r-mer molecules each occupying r sites of a quasi-crystalline lattice. The solution is divided into successive layers and it is supposed that only the first (i.e. the surface) layer has a composition different from that of the bulk. In addition, it is assumed that the r-mers take up only those configurations which are parallel to the surface and that the solutions are athermal.

The partition functions for the surface (Z^m) and for the bulk (Z^l) are calculated by the use of the Flory–Huggins method, and are similar in general form to the combination of (1.80) and (1.81). However as, in this instance, all the surface molecules are considered, the total partition function for the surface (Z^m) is obtained, rather than just that for the adsorbed molecules (Z^{ads}). A convenient starting equation is that for the Gibbs free energy of a surface (1.18),

$$G^m = A^m - \gamma \mathcal{A} + pV^m. \tag{1.18}$$

This choice leads to chemical potentials defined for constant p and γ rather than V and $\mathcal{A}$ (§1.7). The pV^m term is very small and may be neglected. Combination of (1.18), written for the monolayer, with the equation

$$A^m = -kT \ln Z^m \tag{3.38}$$

gives

$$G^m = -kT \ln Z^m - \gamma \mathcal{A}. \tag{3.39}$$

For the bulk phase

$$G^l = A^l + pV^l \approx A^l \tag{3.40}$$

$$= -kT \ln Z^l.$$

The surface and bulk chemical potentials of both monomers and r-mers may then be calculated since (§1.7),

$$\left.\begin{aligned} \mu_1^m &= \left(\frac{\partial G^m}{\partial N_1^m}\right)_{T,p,\gamma,N_2^m}; \quad \mu_1^l = \left(\frac{\partial G^l}{\partial N_1^l}\right)_{T,p,N_2^l} \\ \mu_2^m &= \left(\frac{\partial G^m}{\partial N_2^m}\right)_{T,p,\gamma,N_1^m}; \quad \mu_2^l = \left(\frac{\partial G^l}{\partial N_2^l}\right)_{T,p,N_1^l} \end{aligned}\right\} \tag{3.41}$$

where superscripts m and l refer as usual to the monolayer and bulk liquid. At adsorption equilibrium

$$\mu_1^m = \mu_1^l \quad \text{and} \quad \mu_2^m = \mu_2^l. \tag{3.42}$$

Appropriate combinations of (3.39) to (3.42) yield the desired expressions. The expressions for Z and the ensuing algebra require too much space to be given here, and the interested reader is referred to Prigogine & Maréchal (1952). The results are as follows:

$$\left.\begin{aligned} \gamma &= \gamma_1^0 + \frac{RT}{a_1}\left\{\ln\frac{\phi_1^m}{\phi_1^l} + \frac{r-1}{r}(\phi_2^m - \phi_2^l)\right\} \\ &= \gamma_2^0 + \frac{RT}{ra_1}\left\{\ln\frac{\phi_2^m}{\phi_2^l} + (r-1)(\phi_2^m - \phi_2^l)\right\}. \end{aligned}\right\} \tag{◀3.43}$$

In (3.43) a_1 is the molar surface area of the monomer and the ϕs are *volume fractions* defined as

$$\phi_1 = \frac{N_1}{N_1 + rN_2}; \quad \phi_2 = \frac{rN_2}{N_1 + rN_2}.$$

r_1, which is the ratio of the sizes of the monomer and r-mer, is often put equal to (molar volume of r-mer)/(molar volume of monomer). The expression for the composition of the monolayer may be obtained from (3.43) and is

$$\frac{(\phi_2^m)^{1/r}}{\phi_1^m} = \frac{(\phi_2^l)^{1/r}}{\phi_1^l} \exp\left\{\frac{a_1(\gamma_1^0 - \gamma_2^0)}{RT}\right\}. \tag{3.44}$$

From (3.44) together with the condition $\phi_1 + \phi_2 = 1$, ϕ_1^m and ϕ_2^m may be calculated for various values of ϕ_1^l and ϕ_2^l. Substitution of the resulting data in (3.43) then yields surface tensions which may be compared with observed values. An example of the use of (3.43) is given by Ono & Kondo (1960), who compare experiment and theory for the system benzene–dibenzyl at 60 °C using $r = 2$ and $a_1/N_A = 0.2894\ \text{nm}^2$ (28.94 Å^2). The results are shown in fig. 3.12; the agreement is obviously good.

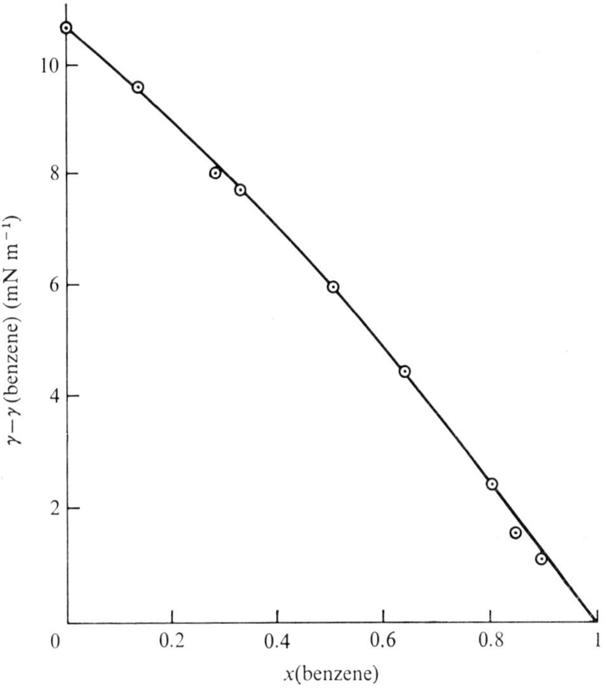

Fig. 3.12. Comparison of experimental and theoretical surface tensions for benzene + dibenzyl mixture at 60 °C. ⊙ are experimental points and the full line is calculated from (3.43) (see text). (After Ono & Kondo, 1960.)

For the special case of $r = 1$, (3.43) reduces to (3.35), the equation for perfect solutions.

3.7. Insoluble monolayers. When placed on a clean water surface, a large number of organic substances, which are insoluble in water, spread and cover the entire surface. The spreading usually occurs so as to produce a monomolecular film. If there is surplus material, this remains as a lens or crystal in equilibrium with the monolayer. It can be demonstrated that the monolayer exerts a two-dimensional (or surface) pressure which is defined as (§1.3)

$$\pi = \gamma_0 - \gamma, \qquad ◀(3.45)$$

where γ_0 and γ are respectively the surface tensions of the pure water and of the water covered by the monolayer. This pressure is simply that which is needed to contain the monolayer in a given area. The thermodynamic significance of the surface pressure is evident from the following consideration. For a surface containing two components, say water and insoluble organic molecules, the excess Helmholtz free energy is given by (1.25) as

$$A^{\sigma} = \gamma \mathcal{A} + \mu_1 n_1^{\sigma} + \mu_2 n_2^{\sigma}. \tag{1.25}$$

If the Gibbs model of the surface is adopted and the surface of division is chosen such that the surface excess of the water is zero ($n_1^{\sigma} = 0$), then

$$A^{\sigma} = \gamma \mathcal{A} + \mu_2 n_2^{\sigma}. \tag{3.46}$$

Although n_2^{σ} is a surface excess quantity, it refers to an insoluble monolayer and is therefore also a total quantity, and could be written n_2^{s}. When no monolayer is present, $n_2^{\sigma} = 0$ and

$$A^{\sigma} = A_0^{\sigma} = \gamma_0 \mathcal{A}. \tag{3.47}$$

Therefore from (3.46) and (3.47)

$$\begin{aligned} A^{\text{monolayer}} &= A^{\sigma} - A_0^{\sigma} \\ &= \gamma \mathcal{A} + \mu_2 n_2^{\sigma} - \gamma_0 \mathcal{A}, \end{aligned}$$

or from (3.45)

$$A^{\text{monolayer}} = -\pi \mathcal{A} + \mu_2 n_2^{\sigma}. \tag{3.48}$$ ◀**(3.48)**

The free energy $A^{\text{monolayer}}$ should be distinguished from A^m in (3.38). $A^{\text{monolayer}}$ reflects only the presence of the insoluble monolayer and becomes zero in absence of the monolayer, whereas A^m reflects equally the presence of both components of the system. $A^{\text{monolayer}}$ is, however, analogous to A^{ads} of §1.10.

For a closed system (n_2^{σ} constant), from (1.15)

$$\frac{\partial A^{\text{monolayer}}}{\partial \mathcal{A}} = \gamma - \gamma_0 = -\pi. \tag{3.49}$$

The surface pressure of an insoluble monolayer is therefore numerically equal to the rate of change of the surface free energy with the area of the monolayer.

The physical significance of the surface pressure of a monolayer at a liquid surface is usually described by analogy with bulk phenomena. Such analogies are never complete and are frequently misleading. The following remarks should therefore be regarded with caution. Liquid surfaces may be thought of as having a wide range of viscosities. Thus, at one extreme the surface may be highly mobile and liquid-like, while at the other it may be very viscous and almost solid (e.g. like a 'glass'). Furthermore, the monolayer molecules may be either almost wholly immersed in the surface layer of the liquid, or they may merely rest on the surface without appreciably penetrating it. For a non-penetrating monolayer or a condensed monolayer (see below) it might be anticipated that the surface pressure would be largely independent of the underlying liquid and, for gaseous monolayers, would be predictable in principle from the equations of §1.11.

For a monolayer which was intimately mixed with the surface layer of the liquid, on the other hand, it would seem more appropriate to regard the surface as a two-dimensional solution (as for the liquid mixtures of §3.6) and to regard the surface pressure as the two-dimensional osmotic pressure of this solution. For gaseous monolayers both approaches have been employed. The osmotic pressure approach is at first sight the most attractive as it more obviously takes account of all molecular species in the surface phase. However, although at a phenomenological level this approach has had some success, it has not contributed greatly to the understanding of the molecular structure of surfaces. This latter criticism is especially true for liquid–liquid interfaces and for ionised monolayers. The gaseous monolayer approach is, by contrast, clearly approximate in that it ignores the presence of the substrate molecules in the surface. For insoluble or strongly adsorbed monolayers, however, it is quite likely, from statistical considerations, that only small errors arise from this neglect. Provided that the solvent is relatively unimportant, the methods of statistical thermodynamics may more readily be applied, so giving a better insight into the behaviour of the monolayer molecules. This approach is especially valuable because it may be used for both liquid–vapour and liquid–liquid interfaces and for monolayers of either non-electrolytes or electrolytes.

The surface pressure of an insoluble monolayer is related through an equation of state to the number of molecules per unit area of the surface. This relationship reflects the molecular state of the monolayer (§1.10) and its experimental determination is usually an early step in any investigation. A large number of reported surface pressures for insoluble monolayers are not equilibrium values. This does not necessarily affect the interrelationship between the surface pressure and area per molecule, but it may complicate the interpretation of experimental data and it is therefore important to consider how the situation arises. For a system in which there are only plane interfaces and in which no chemical reactions occur, the number of degrees of freedom (ω) is given by (Defay, Prigogine, Bellemans & Everett, 1966*e*)

$$\omega = 2+c-\varphi-(\psi-s), \qquad \blacktriangleleft(3.50)$$

where c is the number of components, φ the number of bulk phases, ψ the number of surface phases and s the number of different types of surface.

Suppose that the amount of organic substance added to an interface between pure water and nitrogen is such that it spreads to form a homogeneous monomolecular film, but leaves a crystal or lens in equilibrium on the surface. (It is assumed that the crystal or lens is large compared with molecular dimensions and that its surfaces are therefore effectively planar (§3.2).) Then $c = 3$, $\varphi = 3$, $\psi = s = 3$ and $\omega = 2$. The system has only two degrees of freedom and if the temperature and external pressure are fixed, the composition (or surface pressure) of the monolayer is also fixed. In the presence of excess organic substance there is thus only one equilibrium value of the number of molecules per unit area.

If, on the other hand, the amount of organic substance is such that it all spreads and covers the available surface with a homogeneous film without leaving excess solid, then $c = 3$, $\varphi = 2$, $\psi = s = 1$ and $\omega = 3$. There are now three degrees of freedom and, for constant temperature and external pressure, the number of molecules per unit area may be varied without violating the equilibrium provided it remains below the value which corresponds to the presence of excess solid.

Solid substances at room temperature usually have very small equilibrium values of π. However, both the spreading from the solid to the monolayer at low pressures, and the collapse of the monolayers into bulk solid at high pressures are often very slow processes. As a consequence it is possible in many systems to measure values of π which are far above the equilibrium values, and which correspond to supersaturated monolayers. Stearic acid is a good example of this phenomenon. The equilibrium spreading pressure of solid stearic acid at room temperature is *ca.* 3 mN m^{-1}, but monolayer pressures of more than 30 mN m^{-1} may readily be obtained. The rate of decline of these high pressures to the equilibrium value depends on the circumstances, but may take many hours or even days. In contrast to the situation for solids, the spreading and collapse rates for organic liquids tend to be much more rapid. Thus, for oleic acid, surface pressures of greater than the equilibrium value are not easily attainable and, provided excess acid

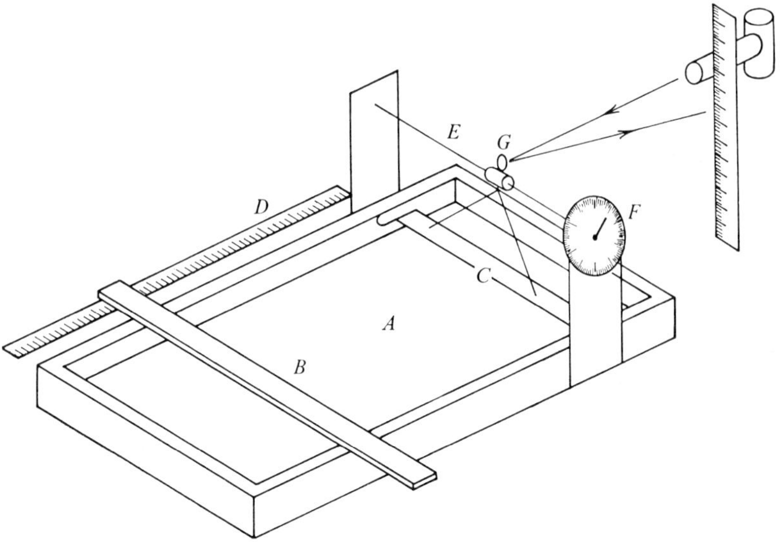

Fig. 3.13. The Langmuir–Adam surface balance (see text).

is present, the system yields a constant surface pressure which is determined only by the temperature and external pressure.

The slow spreading rates of solids, and also the difficulty of weighing out very small quantities of a pure substance, have led to the sometimes dubious practice of forming the monolayer by spreading a dilute solution of the substance in a volatile solvent. The idea is that the solvent evaporates within a few seconds or minutes and leaves the substance spread on the water surface. In practice the solvent often does not entirely disappear, and tends to affect the results.

The surface pressure of an insoluble monolayer may be measured in various ways. The Wilhelmy plate technique described earlier in this chapter may be used to give the tensions of the clean and film-covered surfaces, and thus the surface pressure follows from (3.45). The surface pressure may also be measured directly by means of a Langmuir–Adam surface balance (fig. 3.13). The trough (A) is filled to the brim with water. A known amount of the film-forming substance is spread on the surface of the water such that it is confined by the barrier (B) and the float (C). Thin hydrophobic threads connect the ends of the float to the edges of the trough and prevent leakage of film to the area behind the float. The area of the film is calculated from the positions of the barrier and of the float, which in turn are indicated on the scale (*D*). As the total number of molecules spread on the surface is known, the area per molecule can be found. The surface pressure tends to push back the float (*C*) and is balanced by means of the calibrated torsion wire (*E*). The torsion in the wire, as shown on the scale (*F*) when the float is in its null position, is proportional to the surface pressure. The null position of the float is indicated by means of the light spot reflected from the mirror (*G*). More detailed descriptions of the technique are given in most of the advanced text books on surface chemistry.

Spread films are often not homogeneous. They may contain unspread bulk material or they may be monomolecular but contain more than one surface phase. Inhomogeneity may be detected by various methods, the best known of which are based on optical or surface potential examination. The former method requires the illumination of the film from below by means of a dark ground illuminator, and a means of detecting scattered light, such as a low powered microscope. If the film is monomolecular, no light should be visible but, if there is unspread or collapsed material, this should show up as brightly illuminated regions. The presence of different monomolecular phases is usually indicated by fluctuations in surface potential as an electrode is moved over the surface. The measurement of surface potential has been discussed in §2.3.

Many insoluble films have a considerable resistance to shear in the plane of the film and their two-dimensional shear viscosities and elasticities are readily measurable. The most appropriate technique to use for surface rheological measurements depends very much on the orders of magnitude of the viscous and elastic parameters of the film. There is insufficient space to discuss this question in detail, and it must suffice to say that most approaches are essentially two-dimensional Couette or Poiseuille techniques (Davies & Rideal, 1963).

Insoluble monolayers vary considerably in their properties and a loose system of classification has grown up which is based on the form of the relationship between the surface pressure and area per molecule. There are many substances which, although they have low equilibrium spreading pressures, yield monolayers which are remarkably stable at high surface pressures. In these instances especially, the surface pressure versus area per molecule relationship may be obtained over a wide range of pressure. Some monolayers occupy only small areas per molecule, and are very incompressible. These monolayers are described as *condensed*. At the opposite extreme, a monolayer may occupy a much greater area per molecule, and be very compressible. These monolayers are described as *gaseous*. A variety of intermediate types of behaviour is also encountered.

Insoluble monolayers may exist at oil–water as well as at air–water interfaces. If monolayers of the same substance are compared at the two interfaces, it is usually found that while the one at the air–water interface may be condensed, the one at the oil–water interface is much more expanded and may well be gaseous. The tendency of a monolayer to be more expanded at the oil–water than at the air–water interface appears to be quite general.

Monolayers of non-electrolytes. Many substances with the general formula $C_nH_{2n+1}X$, where X is a polar group, form condensed monolayers. In any given homologous series, the number, n, of carbon atoms necessary for the monolayer to be condensed depends on the nature of the polar group. For un-ionised carboxylic acids and amines and for the alcohols and amides with $n \sim 18$, the monolayers become condensed at room temperature at very low surface pressures. Some examples of the surface pressure versus area per molecule (π versus $\dot{a}$) curves for such substances are given in fig. 3.14. A curve for the steroid cholesterol, which is a particularly good example of a condensed monolayer, has also been included. Owing to the fact that at high pressures such films are often not at equilibrium with small crystalline aggregates, the form of the π versus $\dot{a}$ curve may depend on the time taken to compress the film. Thus, at constant pressure, the area per molecule may appear to decrease with time owing to the partial collapse of the monolayer. The magnitude of this effect varies considerably with the nature of the monolayer, however, and it is very often possible to determine the π versus $\dot{a}$ curve before appreciable collapse occurs. Above a certain range of surface pressure the rate of monolayer collapse increases very rapidly. It is nevertheless difficult to speak of a critical collapse pressure, as this depends very much on the circumstances under which the experiment is carried out.

The concept of the *limiting area per molecule* is one that is often useful, although it is difficult to define in a wholly satisfactory manner. The problem is illustrated in fig. 3.15. If the high pressure part of the π versus $\dot{a}$ curve is linear, it is possible to extrapolate to $\pi = 0$ and so obtain a limiting area per molecule at zero pressure ($\dot{a}_0''$ in fig. 3.15). However, the upper part of the curve is frequently not linear thus ruling out this approach. It may, of course, be possible to extrapolate a lower

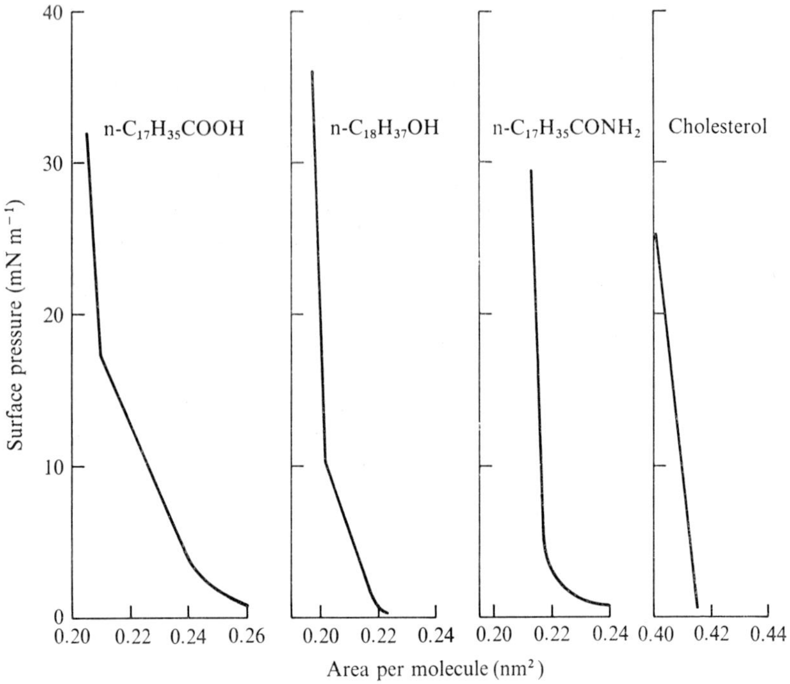

Fig. 3.14. Surface pressure versus area per molecule relationships for some condensed monolayers. The stearic acid was on 0·01 M HCl, the others on water. (The results for stearic acid and stearamide are from Adam (1922), and for cholesterol from Adam & Rosenheim (1929).)

part of the curve in a similar way, so giving another limiting area ($\dot{a}_0'''$). Thirdly, some finite pressure may be chosen and the limiting area defined as the area per molecule at this pressure ($\dot{a}_0'$). These various limiting areas obviously have differing significance and would be used for different purposes. However, it can be seen from fig. 3.14 that there may be little difference between $\dot{a}_0'$ and $\dot{a}_0''$.

Studies of insoluble condensed monolayers revealed at an early stage that within a homologous series of normal alkyl chain substances, the limiting area, $\dot{a}_0$, does not depend on the chain length, and it has been concluded that at high surface pressures, at least, the molecules are oriented with their chains roughly normal to the surface. There are many arguments to suggest that the polar group of the long chain molecules must be in contact with the water rather than on the

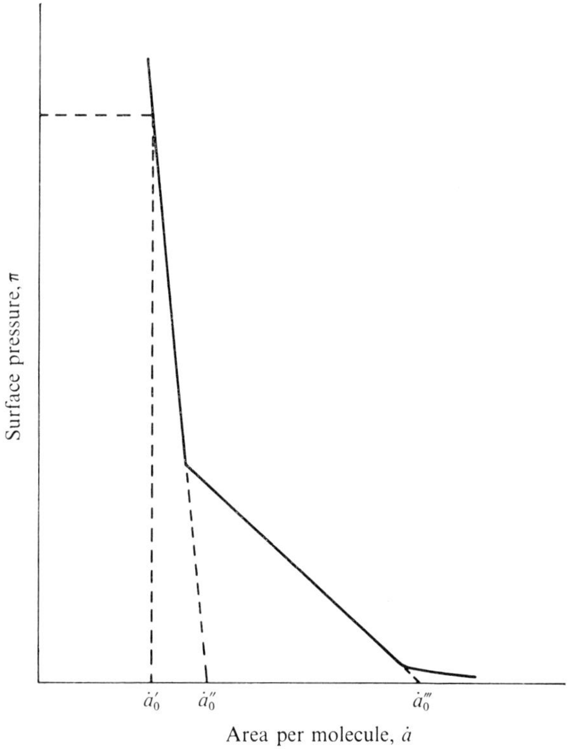

Fig. 3.15. Some definitions of the limiting area per molecule ($\dot{a}_0$) for insoluble monolayers.

upper or air side of the monolayer. One such argument was given earlier (§ 3.4).

The limiting area per molecule for saturated normal alkyl chain substances is approximately constant when the polar group is small. Thus, for the carboxylic acids, amines, amides, methyl ketones, alcohols and some others, the limiting area corresponding to $\dot{a}_0''$ in fig. 3.15 is between 0.20 and 0.22 nm² (20 and 22 Å²) per molecule. It seems probable, therefore, that the value of $\dot{a}_0''$ is governed mainly by the dimensions of the hydrocarbon chain. For substances with a larger polar group, $\dot{a}_0''$ is larger. For the monoglycerides, for example, $\dot{a}_0''$ is about 0.26 nm² per molecule. It will be noted from fig. 3.14 that the curves for the normal chain carboxylic acids and alcohols have a well-defined region of relatively small slope at low pressures. Various

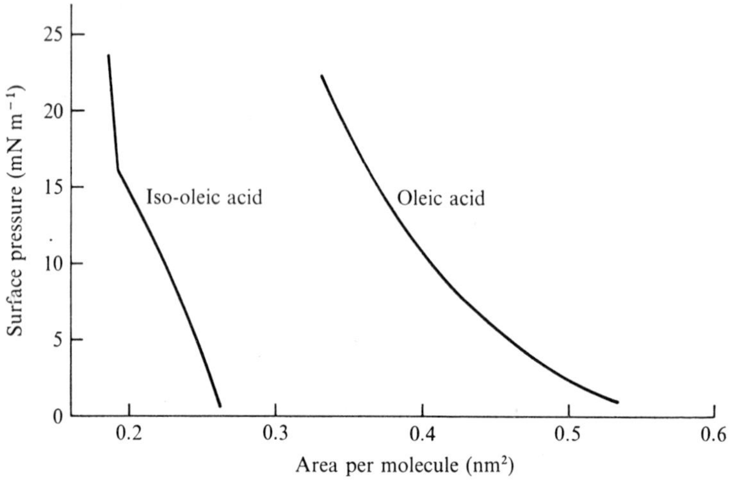

Fig. 3.16. Surface pressure versus area per molecule relationships for oleic (9-octadecenoic) acid and iso-oleic (2-octadecenoic) acid on 0.005 M H_2SO_4 at 15 °C. (After Hughes & Rideal, 1933.)

explanations have been offered to account for this effect but, as yet, there is no very clear evidence as to which is correct.

Monolayers which are intermediate between condensed and gaseous are referred to as *expanded.* Factors which interfere with, or compete with, the tendency of chains to pack closely, promote the formation of expanded monolayers. Thus side chains interfere with packing and lessen inter-chain cohesion. The presence of a second polar group, or even a double bond, introduces another point in the molecule with a strong affinity for the water surface and also tends to disrupt packing. Oleic acid, which has a double bond approximately half way along its chain, is a good example of this latter type of molecule. If the double bond is situated adjacent to the carboxyl group as in iso-oleic acid, the monolayer is condensed (fig. 3.16). No entirely satisfactory quantitative treatment of the π versus $\dot{a}$ curves of either condensed or expanded monolayers has yet been developed, although several attempts have been made for the latter (see e.g. Smith, 1967).

In the discussion of condensed and expanded monolayers the very important influence of temperature has not so far been mentioned. The normal chain saturated carboxylic acids referred to in connection with condensed monolayers, become expanded above a certain temperature or, at a given temperature, below a certain chain length. Thus, increasing

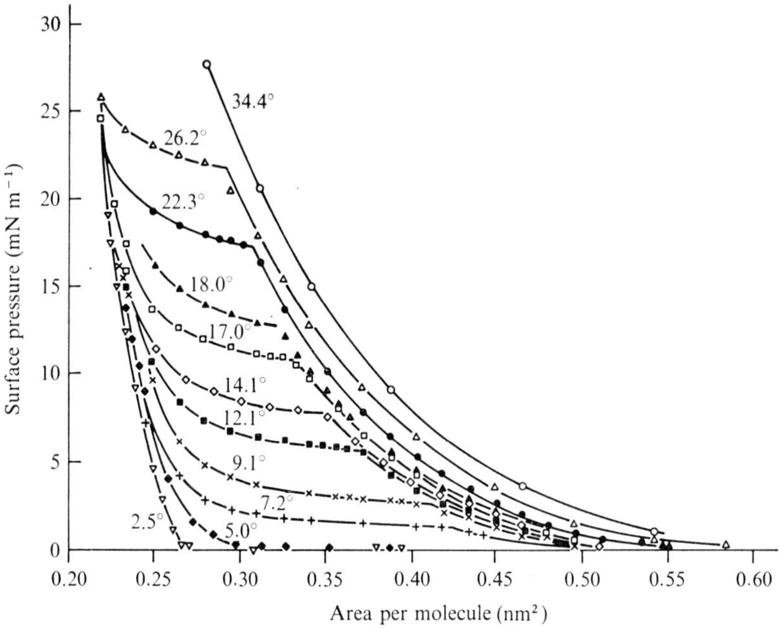

Fig. 3.17. Surface pressure versus area per molecule relationships for myristic acid (n-$C_{13}H_{27}COOH$) on 0.01 M HCl at various temperatures. (After Adam & Jessop, 1926*b*.)

temperature and decreasing chain length both tend to reduce the cohesion between the chains. The temperature effect is well illustrated by the experiments of N. K. Adam on myristic acid (n-$C_{13}H_{27}COOH$) (fig. 3.17). At 2.5 °C the monolayer is almost wholly condensed. As the temperature is raised an expanded region appears at low pressures and large areas per molecule, and this region spreads until eventually the monolayer is in the expanded state at all pressures. At intermediate temperatures there is a well-defined pressure or area per molecule at which the expanded monolayer commences its condensation. The expanded monolayer shows a transition to a gaseous monolayer at a very low surface pressure, as will be mentioned shortly.

The same general factors that cause condensed monolayers to become expanded may, in extreme form, reduce the inter-chain cohesion to such an extent that the monolayers become gaseous. Myristic acid, as just stated, becomes gaseous only at very low pressures at room temperature, and at such a temperature the gaseous region for the longer chain acids is scarcely detectable. For shorter chain lengths, however, the gaseous region becomes predominant. Fig. 3.18 shows the situation for the n-aliphatic carboxylic acids lauric ($C_{11}H_{23}COOH$) to palmitic ($C_{15}H_{31}COOH$). The surface pressures are very low and the areas per molecule are very large. The hyperbolic curves on the right of the figure are characteristic of gaseous monolayers. At a well-defined pressure, condensation occurs for all the substances except lauric acid. Thus, over the horizontal regions of the π versus $\dot{a}$ curves, gaseous and condensed monolayers are in equilibrium with each other. If the phase rule is applied in this region, it is found that $c = 3$, $\varphi = 2$, $\psi = 2$ and $s = 1$, and hence $\omega = 2$. Provided, therefore, that the temperature and external pressure are fixed, the system is invariant, i.e. it can have only one surface pressure, and for this pressure the areas per molecule of the acid in the two surface phases are determined. At 14.5 °C there are *ca.* 0.5 nm² (50 Å²) and 8.5 nm² (850 Å²) per molecule in the liquid and gaseous states respectively. The close similarity between fig. 3.18 and that for the pressure–volume relationships during the condensation of a vapour is obvious. The only difference, other than the units, is that in fig. 3.18 the chain length of the substance and not the temperature has been varied. It has been proved by surface potential measurements that over the horizontal parts of the curves there are in fact two distinct phases present.

With chain lengths of less than twelve carbon atoms the aliphatic carboxylic acids yield gaseous monolayers up to considerably higher surface pressures than those in fig. 3.18. However, such substances are appreciably soluble in water and cannot be treated as insoluble monolayers. Such systems will be considered in §3.8. There are nevertheless ways other than decreasing the chain length, of producing high pressure gaseous monolayers while at the same time maintaining poor solubility of the molecules. One is through the introduction of at least one other polar group in the molecule, at some point well removed from the first. Another is through the introduction of an ionising group into the molecule. The electrostatic repulsion between adjacent ions is then usually sufficient to produce gaseous monolayers, even at high surface pressures.

The ability of two polar groups to produce gaseous monolayers is well exemplified by the dibasic carboxylic esters. The π versus $\dot{a}$ curve for one of these substances is given in fig. 3.19. Neither this system nor the lauric acid in fig. 3.18 obey the equation of state (1.57) for a two-dimensional perfect gas

$$\pi\dot{a} = kT$$

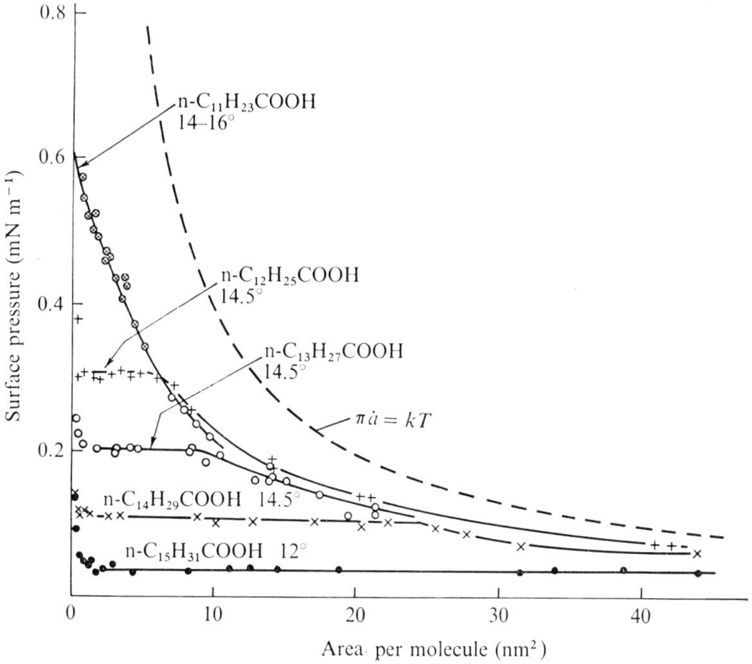

Fig. 3.18. Surface pressure versus area per molecule relationships for a homologous series of aliphatic carboxylic acids. The dashed curve represents the behaviour of an ideal two-dimensional gas at 13 °C. (After Adam & Jessop, 1926*a*.)

although they approach ideality at low pressures. In fact Adam (1968) shows that the correct value of the Boltzmann constant, k, can be obtained from the data of fig. 3.19 as $\pi \rightarrow 0$. The full π versus $\dot{a}$ curves, however, cannot be fitted by the two-dimensional van der Waals equation (1.77). The curve of fig. 3.19 reveals that a well-defined condensation occurs when $\dot{a}$ becomes less than about 0.8 nm^2 (80 Å^2) per molecule, and that the limiting area per molecule in the condensed film so formed is approximately 0.21 nm^2. This suggests that at high surface pressures the dibasic ester is oriented vertically so that only one of the polar groups is in the water. At low pressures, in the gaseous region, the molecules are thought to lie flat on the surface with both their polar regions in the water.

The quantitative interpretation of π versus $\dot{a}$ relationships in terms of the two-dimensional van der Waals equation is usually easier for monolayers at the hydrocarbon–water interface than for those at the

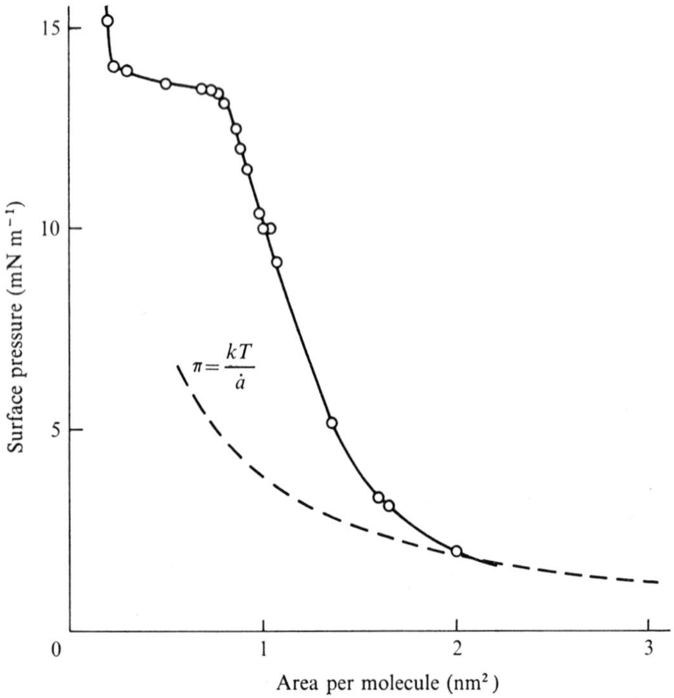

Fig. 3.19. The surface pressure as a function of area per molecule for the di-ester $C_2H_5OOC.(CH_2)_{11}.COOC_2H_5$ at the air–water interface at 1 °C. The dashed curve is for an ideal two-dimensional gas. (After Adam & Jessop, 1926*c*.)

air–water interface. This is because in monolayers at the hydrocarbon–water interface there is frequently little or no inter-chain cohesion, presumably because the energy of interaction between the chains of the monolayer molecules approximates to that between the monolayer molecules and the solvent hydrocarbon. As a consequence, many monolayers which are condensed at the air–water interface are gaseous at a hydrocarbon–water interface. However, most substances which form insoluble monolayers at the air–water interface tend to be soluble in hydrocarbon and can then be examined only by adsorption techniques (§3.8). An important exception to this tendency is found for ionised long chain substances.

Monolayers of electrolytes. The theory of ionised monolayers is very complicated. The description given below is essentially a first approximation but is based on the more rigorous approach of Bell & Levine (1966). A monolayer containing N insoluble molecules in area $\mathcal{A}$ is formed at an interface. N_c of these molecules are assumed to ionise to give univalent ions such that the insoluble ion has charge e^- and the other ion, which is assumed to go into solution, has charge $-e^-$. The free energy of the monolayer is given by an expression analogous to (3.48),

$$A^{\text{monolayer}} = -\pi\mathcal{A} + \sum_i \mu_i N_i^\sigma$$
$$= -\pi\mathcal{A} + \mu_u(N-N_c) + (\tilde{\mu}_c + \mu_s)N_c, \quad (3.51)$$

where $\tilde{\mu}_c$ is the electrochemical potential of a monolayer ion and μ_s and μ_u are respectively the chemical potentials of a soluble ion (as in bulk solution) and an undissociated molecule. The chemical potentials can be shown to be defined by an expression analogous to (1.28),

$$\mu_i = \left(\frac{\partial A^{\text{monolayer}}}{\partial N_i^\sigma}\right)_{T,V^\sigma,N_j^\sigma,a}. \quad (3.52)$$

The change in free energy ($\Delta\tilde{\mu}$) due to the dissociation of one monolayer molecule is given by

$$\Delta\tilde{\mu} = \tilde{\mu}_c + \mu_s - \mu_u \quad (3.53)$$

and in the equilibrium monolayer

$$\Delta\tilde{\mu} = 0. \quad (3.54)$$

From (3.53) and (3.54), (3.51) for the equilibrium state may be rewritten

$$A^{\text{monolayer}} = -\pi\mathcal{A} + \mu_u N. \quad (3.55)$$

It follows from (3.52) that

$$A^{\text{monolayer}} = \int_{N=N_c=0}^{N=N,\, N_c=N_c} \mu_u \,\mathrm{d}(N-N_c) + \tilde{\mu}_c \,\mathrm{d}N_c + \mu_s \,\mathrm{d}N_c \quad (3.56)$$

or, substituting from (3.53) for μ_s,

$$A^{\text{monolayer}} = \int_{N=N_c=0}^{N=N,\, N_c=N_c} \mu_u \,\mathrm{d}N + \Delta\tilde{\mu}\,\mathrm{d}N_c. \quad (3.57)$$

The combination of (3.57) and (3.55) gives

$$\pi\mathcal{A} = \mu_u N - \int_{N=N_c=0}^{N=N,\, N_c=N_c} \mu_u \,\mathrm{d}N + \Delta\tilde{\mu}\,\mathrm{d}N_c. \quad (3.58)$$

On integration by parts and employing (3.54), (3.58) becomes

$$\pi \mathscr{A} = \int_{N=N_c=0}^{N=N,\, N_c=N_c} N\,\mathrm{d}\mu_u + N_c\,\mathrm{d}(\Delta\tilde{\mu}). \tag{3.59}$$

If $\Delta\tilde{\mu}$ is split into its chemical and electrical parts as in (2.4), π may be written formally in terms of electrical and non-electrical contributions. In order to obtain an explicit expression for π, however, it is necessary to choose a partition function for the ionised monolayer. This can be done essentially as in §1.10 and §1.11, except that terms must be included for both charged and uncharged species. In addition the potential energy, V, in (1.52) must include the electrical potential energy $e^-\varphi(0)$ of a monolayer ion at the interface. The non-electrical part of V, denoted V_u, will be assumed as in chapter 1 to be independent of the number, N, of molecules or ions in the surface and, for this reason, does not appear in the equation of state.

The partition function for an ionised monolayer which, apart from its electrical properties, is analogous to the un-ionised monolayers discussed in §1.11 is

$$Z^{\text{monolayer}} = \frac{1}{(N-N_c)!\,N_c!}\left\{\frac{2\pi m_u kT}{h^2} j_u(T)\right\}^{N-N_c}\left\{\frac{2\pi m_c kT}{h^2} j_c(T)\right\}^{N_c}$$
$$\times (\mathscr{A} - N\dot{a}_0)^N \left\{\exp\left(\frac{V_u}{kT}\right)\right\}^N \left\{\exp\left(-\frac{e^-\varphi(0)}{kT}\right)\right\}^{N_c}, \tag{3.60}$$

where m_u and m_c, and $j_u(T)$ and $j_c(T)$ are the masses and internal partition functions of the neutral and ionised monolayer molecules respectively, and $\varphi(0)$ is the mean electrical potential in the plane of the surface ions relative to that in the bulk phase. The quantity $\dot{a}_0$ is the area effectively occupied by a molecule (see footnote, p. 23) and is assumed, together with V_u, to be the same for ionised and un-ionised molecules. It has also been assumed that there are no counter-ions in the plane of the insoluble molecules at the surface. This assumption is probably unreasonable, but the alternatives lead to extremely complicated expressions.

The terms required for (3.59) are obtained from (3.60) by means of the equations

$$A^{\text{monolayer}} = -kT \ln Z^{\text{monolayer}}, \tag{3.61}$$

$$\mu_u = \left(\frac{\partial A^{\text{monolayer}}}{\partial (N-N_c)}\right)_{T,V,\mathscr{A},N_c} \tag{3.62}$$

and

$$\tilde{\mu}_c = \left(\frac{\partial A^{\text{monolayer}}}{\partial N_c}\right)_{T,V,\mathscr{A},(N-N_c)}. \tag{3.63}$$

By differentiation, substitution in (3.59) and integration, the following equation of state is obtained:

$$\pi = \frac{kT}{\dot{a}-\dot{a}_0} + \int_{N_c=0}^{N_c=N_c} \frac{N_c e^-}{\mathcal{A}}\, \mathrm{d}\varphi(0), \tag{3.64}$$

where $\dot{a}$ is the area per ion or molecule, $\mathcal{A}/N$. It will be noted that μ_s (cf. (3.53)) does not appear in (3.64). This is because the monolayer is assumed to be formed on an electrolyte solution which contains the species of soluble ion produced on ionisation of the monolayer. The soluble ion is therefore always in equilibrium with the bulk phase and thus μ_s is always equal to the bulk chemical potential. The latter does not change on formation of the monolayer and therefore μ_s, which is included in $\Delta\tilde{\mu}$ (3.53), vanishes on the differentiation of $\Delta\tilde{\mu}$.

For the final stage of the derivation a relationship between N_c and $\varphi(0)$ is required. The simplest general expression of this type is the Gouy–Chapman equation. If the electrolyte in the substrate is uni-univalent, (2.31) may be used with $z = 1$; i.e.

$$\sigma = \frac{N_c e^-}{\mathcal{A}} = \frac{e^-}{\dot{a}} = (8N(\infty)\epsilon_r\epsilon_0 kT)^{\frac{1}{2}} \sinh\frac{e^-\varphi(0)}{2kT}. \tag{3.65}$$

To be consistent with (3.60) the electrical potential in the bulk phase, $\varphi(\infty)$, has arbitrarily been taken as zero. On substitution for $N_c e^-/\mathcal{A}$ in (3.64) and integration, the result is

$$\pi = \frac{kT}{\dot{a}-\dot{a}_0} + \frac{2kT}{e^-}(8N(\infty)\epsilon_r\epsilon_0 kT)^{\frac{1}{2}}\left\{\cosh\frac{e^-\varphi(0)}{2kT} - 1\right\} \tag{3.66}$$

or, in terms of the surface charge density, σ,

$$\pi = \frac{kT}{\dot{a}-\dot{a}_0} + \frac{2kT}{e^-}(8N(\infty)\epsilon_r\epsilon_0 kT)^{\frac{1}{2}}\left\{\left(\frac{\sigma^2}{8N(\infty)\epsilon_r\epsilon_0 kT} + 1\right)^{\frac{1}{2}} - 1\right\}. \tag{3.67}$$

If $\varphi(0) \gtrsim 100$ mV, i.e.

$$\sinh\frac{e^-\varphi(0)}{2kT} \approx \tfrac{1}{2}\exp\frac{e^-\varphi(0)}{2kT}$$

then (3.67) simplifies to

$$\pi = \frac{kT}{\dot{a}-\dot{a}_0} + \frac{2kT}{\dot{a}_c}, \tag{3.68}$$

where $\dot{a}_c = N_c/\mathcal{A}$. In a monolayer which is almost completely ionised ($N_c \approx N$) the still simpler result

$$\pi = \frac{kT}{\dot{a}-\dot{a}_0} + \frac{2kT}{\dot{a}} \qquad \blacktriangleleft\textbf{(3.69)}$$

is obtained.

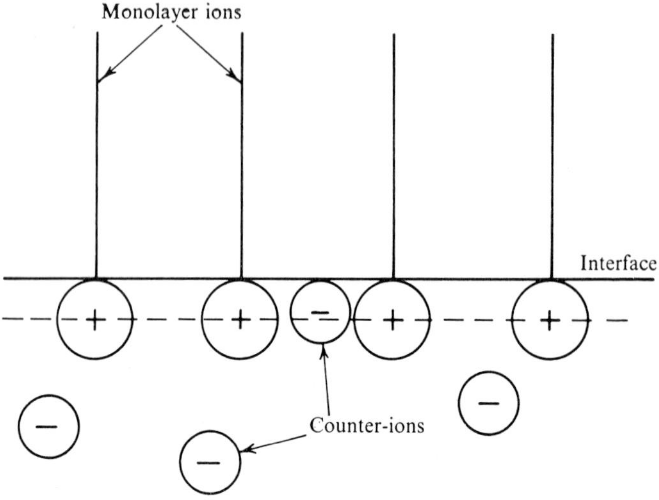

Fig. 3.20. The penetration of counter-ions into and beyond the plane of the monolayer ions.

The validity of (3.66)–(3.69) is, of course, subject to the validity of the Gouy–Chapman equation in this type of system and also to the validity of the partition function (3.60). Some shortcomings of the Gouy–Chapman theory have been discussed in §2.6. In particular it should be stressed that both the discrete nature and the finite size of the ions at the surface have been ignored. It has also been tacitly assumed in the use of (3.65) that counter-ions may not penetrate beyond the plane of the centres of the ionic heads of the monolayer molecules. This assumption is not very reasonable especially for systems in which the ionic head of the monolayer molecule is larger than a counter-ion (fig. 3.20). The partition function does not take account of the presence of counter-ions in the plane of the monolayer ions. This has already been mentioned. A more serious limitation of the partition function, however, is that it takes no account of non-electrostatic interactions between adjacent monolayer molecules. For this reason (3.66)–(3.69) are likely to be more applicable to monolayers at hydrocarbon–water interfaces where interactions between non-polar parts of the molecules are much reduced.

A test of (3.67) is shown in fig. 3.21. Surface pressures of insoluble monolayers of octadecyl trimethylammonium bromide at the n-heptane–0.1 M NaCl interface are compared with the surface pressures

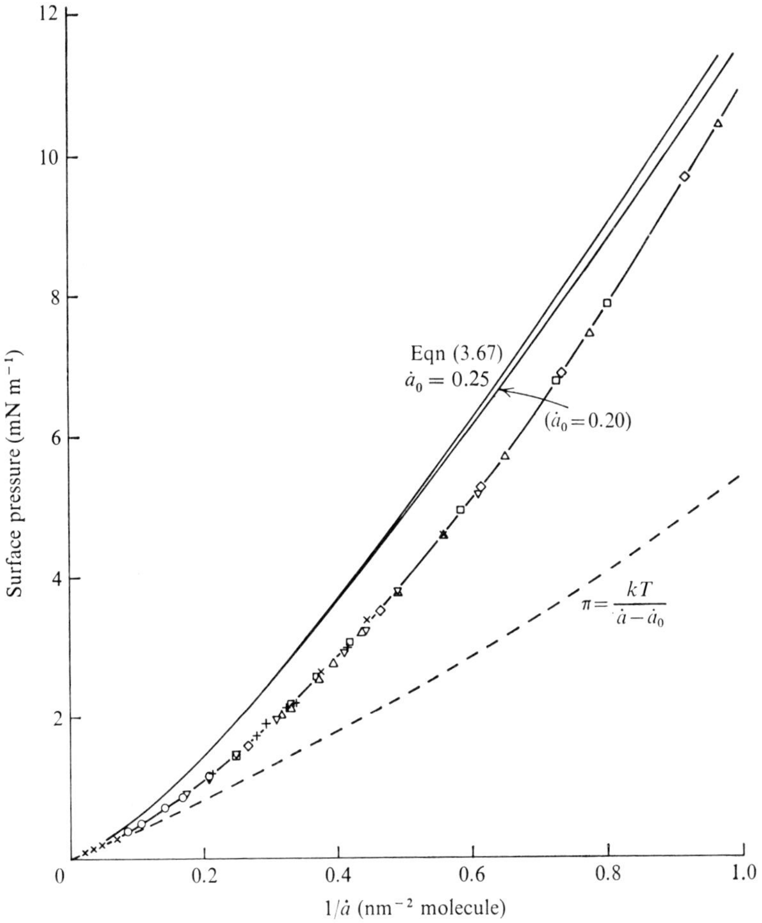

Fig. 3.21. A test of the equation of state (3.67) for an ionised monolayer. The data is for n-octadecyl trimethylammonium bromide spread from a solution in water at the n-heptane–0.1 M NaCl interface at 20 °C. The various symbols represent different experiments. Theoretical pressures, according to (3.67), are shown for two values of $\dot{a}_0$ and are clearly higher than those observed. The dashed curve is a plot of the Volmer equation (1.68) for $\dot{a}_0 = 0.25$ nm^2 molecule^{-1}. (Unpublished results of J. Mingins and G. Taylor.)

predicted from (3.67). There is a serious disagreement at all but very low pressures and, although there is some uncertainty in the appropriate value of $\dot{a}_0$ to use in (3.67), it is not likely that this could account for the whole discrepancy. The surface pressure of an un-ionised monolayer as predicted by the Volmer equation (1.68) is included for comparison.

3.8. Adsorption from dilute solutions. Dilute solutions nearly always have surface or interfacial tensions which differ from that of the pure solvent. If the tension decreases as the solute concentration increases, the Gibbs equation (1.43) shows that the surface excess of the solute is positive and that adsorption occurs. This is the situation encountered for aqueous solutions of many organic solutes. For some inorganic electrolytes in water, however, the tension increases with concentration. In these instances the surface excess of the electrolyte is negative and desorption occurs.

The adsorbed solute almost invariably takes the form of a uniform monomolecular layer similar to those described in the previous two sections. The surface excess may be determined by means of the Gibbs equation

$$\Gamma_2^{(1)} = -\frac{1}{RT}\frac{\mathrm{d}\gamma}{\mathrm{d}\ln \boldsymbol{a}_2} \tag{1.43}$$

provided that the surface and bulk of the system are in equilibrium. The interfacial tension must be measured as a function of the solute activity at constant temperature. Various procedures may then be used. Probably the most common is to plot the tension against the logarithm of the solute activity and to obtain $\mathrm{d}\gamma/\mathrm{d}\ln \boldsymbol{a}_2$ graphically. Alternatively, the experimental data may be fitted with a polynomial and the gradients obtained by differentiation. Whichever method is adopted, very precise experimental results are required if the surface excess is to be reasonably accurate.

Considerable effort has been made to verify the Gibbs equation. Many attempts have not succeeded and of those that have, most have been rather inaccurate. A successful method was devised by McBain and his colleagues (McBain & Humphreys, 1932; McBain & Swain, 1936) which involved the removal and collection of a very thin (*ca.* 0.1 mm) layer of liquid from the surface of a solution. From the difference in concentration between this surface liquid and the bulk solution, the surface excess was calculated. The results for aqueous solutions of *p*-toluidine, phenol, n-hexanoic acid and sodium chloride coincided within experimental error with those obtained from the application of the Gibbs equation to surface tension data. More recently, isotopically labelled molecules have been employed in the direct measurement of adsorption. This method has been applied chiefly to the adsorption of surface active ions. Here again, the results are generally not very accurate but they do appear to confirm the validity of the Gibbs equation. A more detailed description of this work is given by Adamson (1967*c*).

Non-electrolytes. Although most non-electrolytes are adsorbed from dilute aqueous solution, relatively few have been studied systematically. A large amount of work has, however, been carried out on the normal chain paraffinic compounds having a polar group in the '1' position. These substances are strongly adsorbed from aqueous solutions to both air– and oil–water interfaces and many of them are of great importance in technology. The basic principles of their behaviour are well illustrated by the n-alkanols.

The adsorption of alkanols is less complicated at hydrocarbon–water than at air–water interfaces and for this reason the former systems will be described first.

The general form of the Gibbs equation for systems which comprise an oil–water interface and a solute which is soluble in both oil and aqueous phases is quite complicated (Moilliet, Collie & Black, 1961). However, provided the oil and water are only very sparingly mutually soluble, an equation similar to (1.43) may be used. If also the solute is strongly adsorbed relative to the solvents, the surface excess may be equated to the surface concentration with negligible error and hence the area per molecule, $\dot{a}$, of the solute may be obtained. At low concentrations, n-alkanols in aliphatic hydrocarbon–water systems satisfy these conditions. The model which accounts very well for the data is that for non-localised monolayers with negligible intermolecular cohesion (§1.11). The Volmer equation of state

$$\pi = \frac{kT}{\dot{a}-\dot{a}_0} \tag{1.68}$$

is tested in fig. 3.22 by plotting $1/\pi$ against $\dot{a}$. The plot is linear and of the correct slope ($1/kT$). The intercept gives $\dot{a}_0 = 0.24$ nm^2 per molecule, which is somewhat similar to that expected from the cross-sectional area, taken normal to the long axis, of an extended n-octanol molecule (see §1.11).

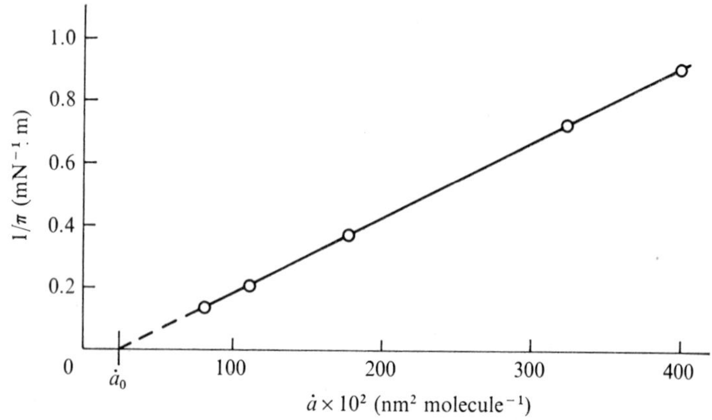

Fig. 3.22. A test of the Volmer equation of state (1.68) for n-octanol adsorbed at the n-dodecane–water interface. The extrapolation of the plot to $1/\pi = 0$ gives $\dot{a}_0 = 0.24$ nm² (24 Å²) molecule⁻¹. $T = 25$ °C. (Data taken from Aveyard & Briscoe, 1970.)

The adsorption isotherm which corresponds to (1.68) is essentially (1.75), although it is necessary to adapt this equation for adsorption from solution rather than from the gas phase. This may readily be achieved by the use of Henry's law. At equilibrium, solute vapour will be in equilibrium with solute in the bulk liquid phases and with solute at the interface. It is therefore permissible, for ideal systems, to write

$$p = \beta x, \tag{3.70}$$

where p is the partial pressure of the solute in the vapour phase, x is its mole fraction in solution and β is the Henry's law constant. Equation (1.75) may then be combined with (3.70) to give

$$x = \frac{K_1}{\beta}\frac{\dot{a}_0}{\dot{a}-\dot{a}_0}\exp\frac{\dot{a}_0}{\dot{a}-\dot{a}_0}. \tag{3.71}$$

Again using (3.70), (1.100) may be rewritten

$$\Delta_a\mu = RT\ln\frac{x}{x^{\ominus}}. \tag{3.72}$$

Substituting for x in (3.72) and invoking the standard states $\theta = \dot{a}_0/\dot{a} = \frac{1}{2}$ and $x^{\ominus} = 1$ (but where the behaviour is identical to that in ideal dilute solution), it is found that

$$\Delta_a\mu^{\ominus} = RT\ln\frac{K_1}{\beta}+RT \tag{3.73}$$

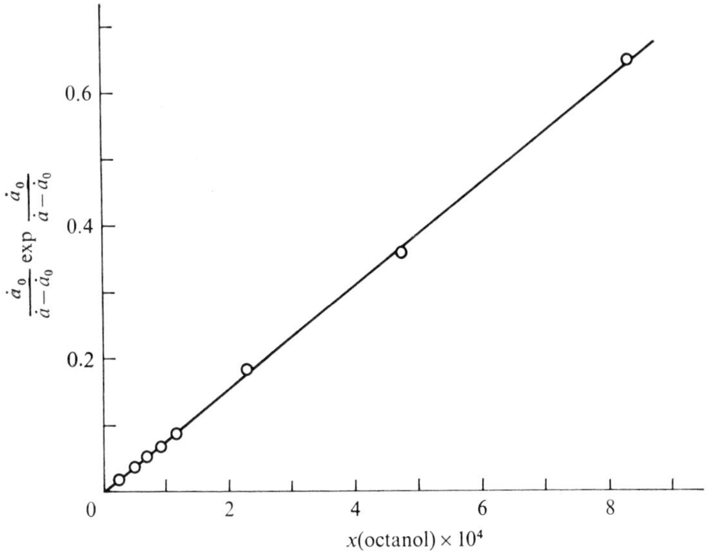

Fig. 3.23. A test of (3.74) for n-octanol adsorbed at the n-dodecane–water interface. The value of $\dot{a}_0$ was taken as 0.24 nm² molecule⁻¹. $T = 25$ °C. (Data taken from Aveyard & Briscoe, 1970.)

and hence, for adsorption from ideal solution,

$$x = \frac{\dot{a}_0}{\dot{a}-\dot{a}_0} \exp \frac{\dot{a}_0}{\dot{a}-\dot{a}_0} \exp \left(\frac{\Delta_a \mu^\ominus}{RT} - 1\right). \qquad \blacktriangleleft(\mathbf{3.74})$$

A plot of the adsorption isotherm (3.74), for adsorption from ideal dilute solutions of n-octanol in dodecane at 25 °C, is shown in fig. 3.23. The value of $\dot{a}_0$ was obtained from fig. 3.22. As would be expected from the linear plot for the equation of state, the plot of the isotherm is also linear.

Several of the n-alkanols have been examined in this way, the higher ones by adsorption from the hydrocarbon phase and the lower members by adsorption from the aqueous phase. For each system the standard free energy of adsorption, $\Delta_a \mu^\ominus$, has been calculated from the slope of the isotherm. For adsorption from hydrocarbon it is found that $\Delta_a \mu^\ominus$ is almost independent of the chain length of the alkanol. This is consistent with the idea that in the process of adsorption, the portion of the alkanol molecule which transfers to the aqueous phase is always the same and most probably is just the hydroxyl group. For adsorption from the aqueous phase, on the other hand, $\Delta_a \mu^\ominus$ is strongly dependent

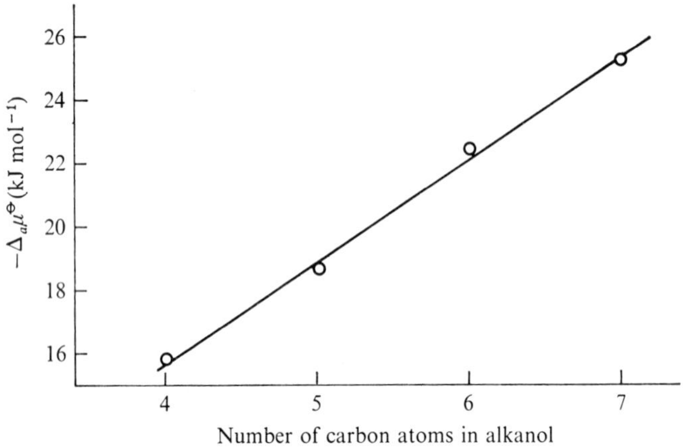

Fig. 3.24. The standard free energy of adsorption $\Delta_a\mu^\ominus$ for the n-alkanols from water to the n-dodecane–water interface. $\Delta_a\mu^\ominus$ was calculated from plots of (3.74). $T = 20$ °C. (Data taken from Mitchell, 1969.)

on the chain length of the alkanol (fig. 3.24). The slope of this plot gives the free energy of adsorption per methylene group. This is approximately constant and has the value -3.2 kJ mol^{-1} at 20 °C. From the distribution of n-alkanols between water and aliphatic hydrocarbon it is found that the free energy of transference of the methylene group is also approximately constant and of value -3.3 kJ mol^{-1}. The close agreement of these two free energies obviously supports the supposition that on adsorption from water to a hydrocarbon–water interface, the chain of the alkanol transfers to a large extent into the hydrocarbon.

The n-alkanols adsorb almost as strongly at air–water interfaces as at hydrocarbon–water interfaces. The resulting monolayers do not, however, obey a simple equation of state, such as the Volmer equation (1.68). This is probably because the mutual attractions of the hydrocarbon chains are not screened, as they are at the hydrocarbon–water interface. Equations similar to the two-dimensional van der Waals equation (1.77) have been applied to monolayers at the air–water interface, but so far no satisfactory interaction term has been derived for this type of system. The same problems arise in connection with the adsorption isotherm. In order to obtain standard free energies in such markedly non-ideal systems as these, it is necessary to examine the region of very low adsorption where, to a good approximation, the simple equation of state

$$\pi\dot{a} = kT \tag{1.57}$$

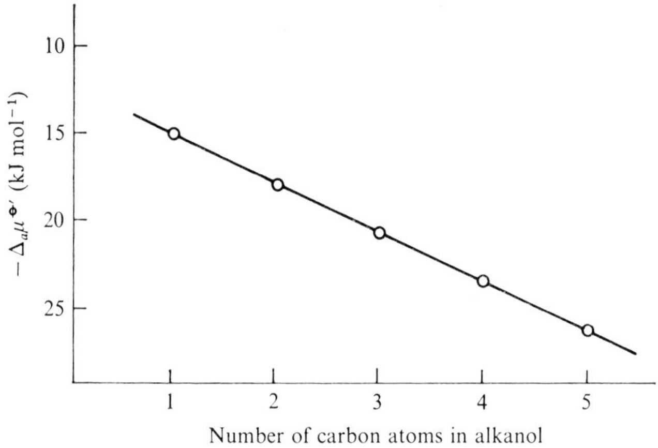

Fig. 3.25. The standard free energy of adsorption $\Delta_a \mu^{\ominus'}$ for the n-alkanols from water to the air–aqueous solution interface. $\Delta_a\mu^{\ominus'}$ was calculated from (3.78). $T = 25$ °C. (After Clint *et al.* 1968.)

is obeyed. It is particularly convenient, under these circumstances, to use an approach to the standard free energies somewhat different from that described above for the hydrocarbon–water systems. By substitution in the Gibbs equation (1.42) for $\Gamma_2^{(1)}(\approx 1/N_A\dot{a})$ from (1.57), equation (3.75) is obtained (Betts & Pethica, 1960):

$$RT\frac{d\pi}{\pi} = d\mu_2. \tag{3.75}$$

By choosing the standard state of the monolayer as unit ideal surface pressure, (3.75) may be integrated to give

$$\mu_2 = \mu_2^{\ominus, s'} + RT \ln \pi. \tag{3.76}$$

If the bulk solution is ideal

$$\mu_2 = \mu_2^{\ominus, l} + RT \ln x_2^l \tag{3.77}$$

and hence the standard free energy of adsorption is, for this alternative choice of standard state in the surface,

$$\Delta_a\mu^{\ominus'} = \mu_2^{\ominus, s'} - \mu_2^{\ominus, l} = RT \ln (x_2^l/\pi). \qquad \blacktriangleleft\textbf{(3.78)}$$

$\Delta_a\mu^{\ominus'}$ is found from the limiting slope of a plot of π against x_2^l. Values of $\Delta_a\mu^{\ominus'}$ for the adsorption of a homologous series of n-alkanols at the air–water interface are shown in fig. 3.25. As for the hydrocarbon–water interface, the plot is linear, but in this instance the slope, or $\Delta_a \mu^{\ominus\prime}$ per $-CH_2-$, is -2.71 kJ mol^{-1} as compared with -3.2 kJ mol^{-1}. This result is consistent with the supposition that at the air–water interface the alkanol chain does not escape entirely from contact with water (it probably lies on the surface) and that at infinite dilution it does not associate with other alkanol chains.

Electrolytes. The adsorption of electrolytes may be determined by means of the Gibbs equation in much the same way as for non-electrolytes. The procedure is, however, not quite straightforward as is shown by the following examples. Consider a solution of a uni-univalent electrolyte A^+B^- in water, in contact with its vapour. Five ionic and molecular species A^+, B^-, H^+, OH^- and H_2O have been brought together in the system, and each must, at the outset, be formally included in the Gibbs equation. Using the surface phase formulation, (1.40) is written

$$-d\gamma = \Gamma^s_{A^+} d\mu_{A^+} + \Gamma^s_{B^-} d\mu_{B^-} + \Gamma^s_{H^+} d\mu_{H^+} + \Gamma^s_{OH^-} d\mu_{OH^-} + \Gamma^s_{H_2O} d\mu_{H_2O}. \tag{3.79}$$

In many instances, but not in very dilute solutions, the terms for H^+ and OH^- may be neglected. Assuming this to be permissible, the combination of (3.79) with the Gibbs–Duhem expression

$$n_{A^+} d\mu_{A^+} + n_{B^-} d\mu_{B^-} + n_{H_2O} d\mu_{H_2O} = 0 \tag{3.80}$$

yields

$$-d\gamma = \left(\Gamma^s_{A^+} - \frac{n_{A^+}}{n_{H_2O}} \Gamma^s_{H_2O}\right) d\mu_{A^+} + \left(\Gamma^s_{B^-} - \frac{n_{B^-}}{n_{H_2O}} \Gamma^s_{H_2O}\right) d\mu_{B^-}. \tag{3.81}$$

The requirement that both bulk and surface phases should be electrically neutral may be expressed

$$n_{A^+} = n_{B^-} = n_{AB} \tag{3.82}$$

and

$$\Gamma^s_{A^+} = \Gamma^s_{B^-} = \Gamma^s_{AB}. \tag{3.83}$$

Equation (3.81) may then be written

$$-\frac{d\gamma}{RT} = \left(\Gamma^s_{AB} - \frac{n_{AB}}{n_{H_2O}} \Gamma^s_{H_2O}\right) d \ln \boldsymbol{a}_{A^+} \boldsymbol{a}_{B^-} \tag{3.84}$$

or, since the mean activity of AB, $\boldsymbol{a}_{\pm}$, is given by

$$\boldsymbol{a}^2_{\pm} = \boldsymbol{a}_{A^+} \boldsymbol{a}_{B^-}, \tag{3.85}$$

$$-\frac{d\gamma}{RT} = 2\left(\Gamma^s_{AB} - \frac{n_{AB}}{n_{H_2O}} \Gamma^s_{H_2O}\right) d \ln \boldsymbol{a}_{\pm}. \tag{3.86}$$

In terms of the surface excess $\Gamma^{(1)}_{AB}$ of AB relative to H_2O

$$-\frac{d\gamma}{RT} = 2\Gamma^{(1)}_{AB}\, d \ln \boldsymbol{a}_{\pm}. \qquad \blacktriangleleft\textbf{(3.87)}$$

The comparison of (3.87) with (1.43) shows that they differ in form by a factor of two. Physically this means that if, in the two different systems, $\Gamma^{(1)}_{AB} = \Gamma^{(1)}_{2}$ and $\boldsymbol{a}_{\pm} = \boldsymbol{a}_2$, the variation of surface tension with activity is twice as large for the completely dissociated 1:1 electrolyte as for the non-electrolyte. This difference in behaviour between electrolytes and non-electrolytes in a surface phase is analogous to the difference between their colligative properties in a bulk phase. Thus the factor of two in (3.87) may be compared with the van't Hoff '*i*' factor of two which occurs in the bulk phase osmotic pressure relationship for dilute solutions of strong 1:1 electrolytes.

In a second example, a system will be chosen for which a factor of two does not appear in the final equation. Consider a solution in water of two completely dissociated uni-univalent electrolytes A^+B^- and M^+B^-. If the dissociation of the water is again negligible, (1.40) becomes

$$-d\gamma = \Gamma^{s}_{A^+}\, d\mu_{A^+} + \Gamma^{s}_{M^+}\, d\mu_{M^+} + \Gamma^{s}_{B^-}\, d\mu_{B^-} + \Gamma^{s}_{H_2O}\, d\mu_{H_2O}. \qquad (3.88)$$

It is quite common in practice for the adsorption of a strongly surface active substance (say A^+B^-) from a dilute solution to be studied in the presence of a large and constant concentration of a much less surface active substance (say M^+B^-). The variation of the concentration of A^+B^- does not, under these conditions, produce an appreciable variation in B^-, and both $d\mu_{M^+}$ and $d\mu_{B^-}$ may be neglected in comparison to $d\mu_{A^+}$. The equation for this system which is comparable to (3.87) is therefore readily derived as

$$-\frac{d\gamma}{RT} = \Gamma^{(1)}_{A^+}\, d \ln \boldsymbol{a}_{A^+} = \Gamma^{(1)}_{AB}\, d \ln \boldsymbol{a}_{A^+}. \qquad (3.89)$$

Now $\boldsymbol{a}_{A^+}$ is a single ion activity and therefore is indeterminate. However, the activity coefficient γ_{A^+} would be mainly determined by the large and constant concentration of M^+B^- and should thus be effectively constant. Equation (3.89), for practical purposes, then becomes

$$-\frac{d\gamma}{RT} = \Gamma^{(1)}_{A^+}\, d \ln m_{A^+} = \Gamma^{(1)}_{AB}\, d \ln m_{AB}, \qquad \blacktriangleleft\textbf{(3.90)}$$

where concentrations are expressed as molalities, *m*. It is a simple

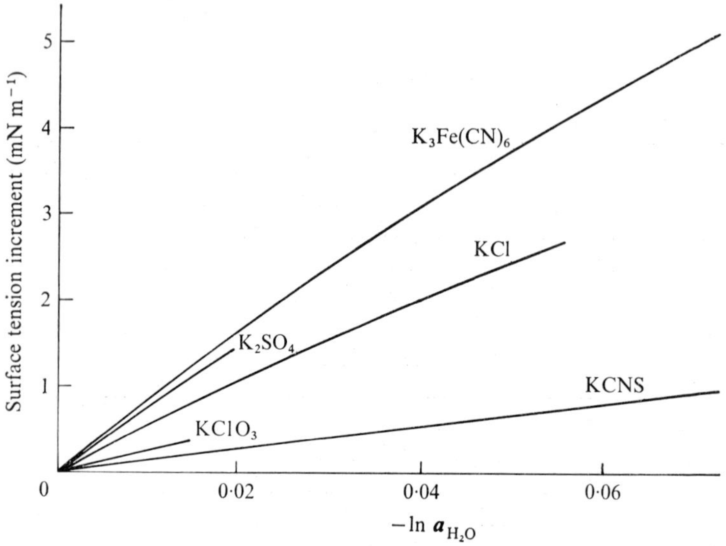

Fig. 3.26. The surface tension, relative to that of pure water, of some electrolyte solutions. The activity a_{H_2O} of the water is related to the molality m of the electrolyte by the expression $-\ln a_{H_2O} = 18\nu m\varphi/1000$, where ν is the number of ions produced by one electrolyte molecule and φ is the molal osmotic coefficient. According to the Gibbs equation the gradient of the curves is proportional to the surface excess of water. (Data of Jones and Ray, after Randles, 1963.)

exercise to derive a general equation applicable to both the above systems. Both (3.87) and (3.89) have been widely used in the study of electrolyte adsorption. Investigations in this field fall largely into two categories. The first deals with inorganic electrolytes which, at most, are only weakly adsorbed, while the second concerns organic ions, especially those with long alkyl chains. The latter are very strongly adsorbed from aqueous solutions and, as is well known, form the basis of many commercial surface active agents.

As mentioned earlier, the surface tensions of many aqueous inorganic electrolyte solutions increase with increasing concentration of the electrolyte. The experimental results for some potassium salts are shown in fig. 3.26. The changes in tension are not large, although for electrolytes such as $CaCl_2$, which are very soluble in water, $\Delta\gamma$ may reach more than 20 mN m^{-1} at high concentrations. The application of the Gibbs equation to these results obviously indicates that the surface excesses of the various salts are negative or, in other words, that the electrolyte as a

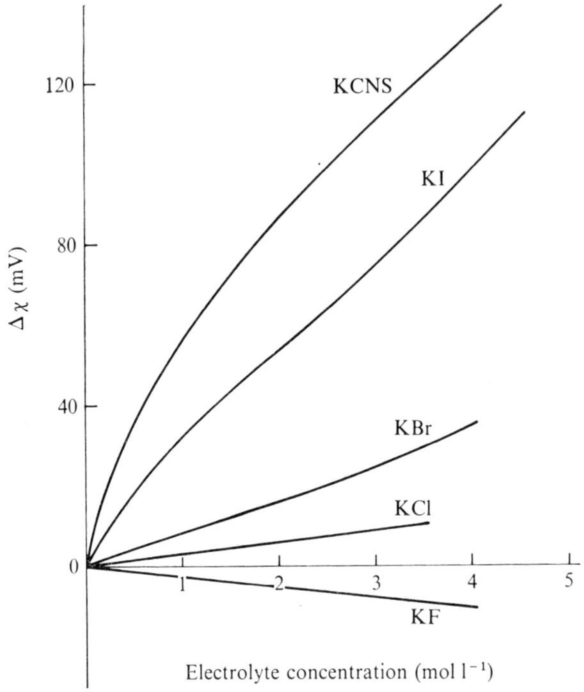

Fig. 3.27. The surface potentials of some aqueous solutions of electrolytes. (After Frumkin, 1924.)

whole is desorbed from the interface. This conclusion reveals nothing concerning the relative behaviour of the positive and negative ions. It may well be that one ion is repelled from the surface more strongly than the other, and for information on this question it is useful to examine the surface potentials. Some data are given in fig. 3.27. $\Delta\chi$, as explained in §§2.2–6, arises from changes in both the dipole orientation at the interface and in the ionic double layer. Owing to the fact that the inorganic electrolytes are desorbed the former contribution is probably small, and $\Delta\chi$ in these systems is usually thought to reflect approximately the change in the ionic double layer potential. The observed $\Delta\chi$ is in most instances positive, i.e. the interior of the solution is positive relative to the surface. It would appear, therefore, that the cations tend to be repelled more strongly than the anions, except in the case of KF. The consideration of the whole range of results for inorganic salts suggests that the small strongly hydrated ions tend to be repelled more strongly than the larger and less strongly hydrated ions. Such differences between the various electrolytes are, however, prominent mainly at fairly high concentrations. In dilute solution, equivalent concentrations of salts of similar valence type have similar effects on surface tension. This observation has led to several attempts to predict the tension increments on the basis of electrostatic models. It would be expected that coulombic forces should contribute to the desorption of the ions. Thus, the elementary electrostatic theory of image forces shows that a charged particle in a medium of high dielectric constant (e.g. water) should be repelled from the surface of a medium of low dielectric constant (e.g. air or hydrocarbon). The development of this approach first by Onsager & Samaras (1934) and then in more detail by others (e.g. Schmutzer, 1955) has led to reasonable agreement between theory and experiment at low or medium concentrations. In addition to the coulombic-cum-hydration contributions to the total interaction with the interface, which are repulsive, there is some evidence for specific attractive forces in certain instances. For example, KPF_6 and some inorganic acids lower the surface tension of water (Randles, 1963). Nothing is known, however, concerning the nature of these forces.

The concluding section of this chapter will be devoted to a brief discussion of the adsorption of normal alkyl chain ions. A substantial volume of the literature is concerned with this subject although it presents serious difficulties, both experimentally and theoretically.

An adsorption isotherm for a strongly adsorbed ion from dilute solution may be obtained in the following way. The partition function for a monolayer of adsorbed ions is that given for an insoluble monolayer by (3.60), and the electrochemical potential of the surface ions is given by (3.63). If the bulk solution of the ions is assumed to be ideal and at zero potential, the electrochemical potential $\tilde{\mu}_c$ in the bulk phase is given by

$$\tilde{\mu}_c = \mu_c^{\ominus, l} + RT \ln x_c. \qquad (3.91)\dagger$$

At equilibrium the electrochemical potentials of the ions in bulk and surface are equal and if, as previously, it is assumed that the monolayer is almost completely ionised ($N_c \approx N$) the isotherm

$$x_c = K_i \frac{\dot{a}_0}{\dot{a} - \dot{a}_0} \exp \frac{\dot{a}_0}{\dot{a} - \dot{a}_0} \exp \frac{e^{-}\varphi(0)}{kT} \qquad (3.92)$$

† The mole fraction has been used as the concentration parameter in this equation in order to maintain the comparability of the standard free energies of adsorption in (3.74) and (3.95)

may be derived, where

$$K_i = \left[\frac{2\pi m_c kT}{h^2}\dot{a}_0 j_c(T) \exp\frac{\mu_c^{\ominus, l}}{RT}\right]^{-1}. \tag{3.93}$$

The relationship between K_i and the standard free energy of adsorption $\Delta_a\mu^{\ominus}$ may be derived in a manner analogous to that for non-electrolytes. The only difference lies in the additional requirement of zero potential for the standard state in the surface. Thus,

$$K_i = \exp\left(\frac{\Delta_a\mu^{\ominus}}{RT} - 1\right), \tag{3.94}$$

where the standard states are, for the surface, ideal half coverage, i.e. $\dot{a} = 2\dot{a}_0$ and $\varphi(0) = 0$, and for the bulk, $x_c = 1$. Equation (3.92) may now be written

$$x_c = \frac{\dot{a}_0}{\dot{a}-\dot{a}_0}\exp\frac{\dot{a}_0}{\dot{a}-\dot{a}_0}\exp\left(\frac{\Delta_a\mu^{\ominus}}{RT}-1\right)\exp\frac{e^-\varphi(0)}{kT}. \quad \blacktriangleleft\textbf{(3.95)}$$

It should be noted that (3.95) differs from the corresponding equation for non-electrolytes (3.74) only in the final exponential term. Thus, as would be expected, the electrical potential of the surface influences the adsorption of the ions. The way in which this occurs is seen more clearly if (3.95) is rearranged in the form,

$$\frac{\dot{a}_0}{\dot{a}-\dot{a}_0}\exp\frac{\dot{a}_0}{\dot{a}-\dot{a}_0} = x_c \exp\left(-\frac{\Delta_a\mu^{\ominus}}{RT}+1\right)\exp\left(-\frac{e^-\varphi(0)}{kT}\right). \tag{3.96}$$

The left-hand side of (3.96) is a function only of $\dot{a}$, and increases with x_c at a rate which is dependent on the two energy terms on the right. The first of these terms,

$$\exp(-\Delta_a\mu^{\ominus}/RT+1)$$

is a constant. The second, $\exp(-e^-\varphi(0)/kT)$,

however, depends on $\dot{a}$ according to (3.65). If the electrolyte concentration $N(\infty)$ is constant, then $\exp(-e^-\varphi(0)/kT)$ decreases with decreasing $\dot{a}$ and, as the adsorption proceeds, the function on the left increases with x_c less and less rapidly. This is in contrast to the situation for non-electrolytes and illustrates the importance of the surface potential in inhibiting adsorption.

The equation of state for adsorbed ions which corresponds to the isotherm (3.95) may be derived by reversing the procedure used in (1.71)–(1.75). Thus, the logarithm is taken of both sides of the isotherm, and the resulting equation is differentiated. Then, x_c may be eliminated by

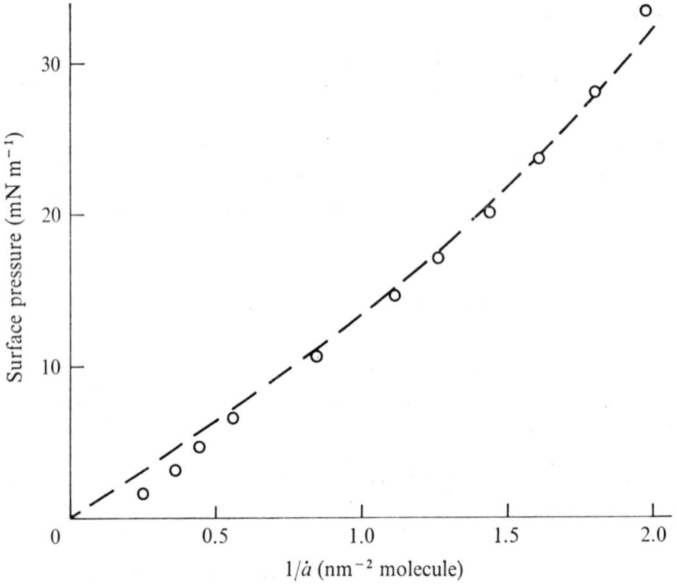

Fig. 3.28. A test of the equation of state (3.67) for an ionized monolayer. The data is for n-dodecyl trimethylammonium bromide adsorbed at the n-decane–5×10^{-3} M NaCl solution interface at 20 °C. Theoretical pressures, according to (3.67), with $\dot{a}_0 = 0.25$ nm^2 molecule^{-1}, are indicated by the dashed curve. (Results of Carroll, 1970.)

means of the Gibbs equation and, on combination with the Gouy–Chapman equation (3.65) and integration, the equation of state is obtained. The procedure is complicated by the fact that the Gibbs equation for electrolytes has different forms under different conditions (cf. (3.87) and (3.89)) and also by the fact that when only the surface active electrolyte is present, the electrolyte concentration $N(\infty)$ in the Gouy–Chapman equation (3.65) is a variable. The resulting equation of state is, however, identical to that for insoluble monolayers (3.67).

As for insoluble monolayers, tests of (3.67) and of (3.95) are rendered difficult by the absence of a satisfactory method for the determination of $\dot{a}_0$. Owing to the electrostatic interaction of the ions there is no plot comparable to that shown in fig. 3.22 for non-electrolytes, which may be used for this purpose. As a consequence, $\dot{a}_0$ has to be estimated from molecular models. Although this may incur little error at low surface pressures where $\dot{a} \gg \dot{a}_0$, it may be a serious source of error at high pressures. There is also a scarcity of suitable adsorption data. Owing to

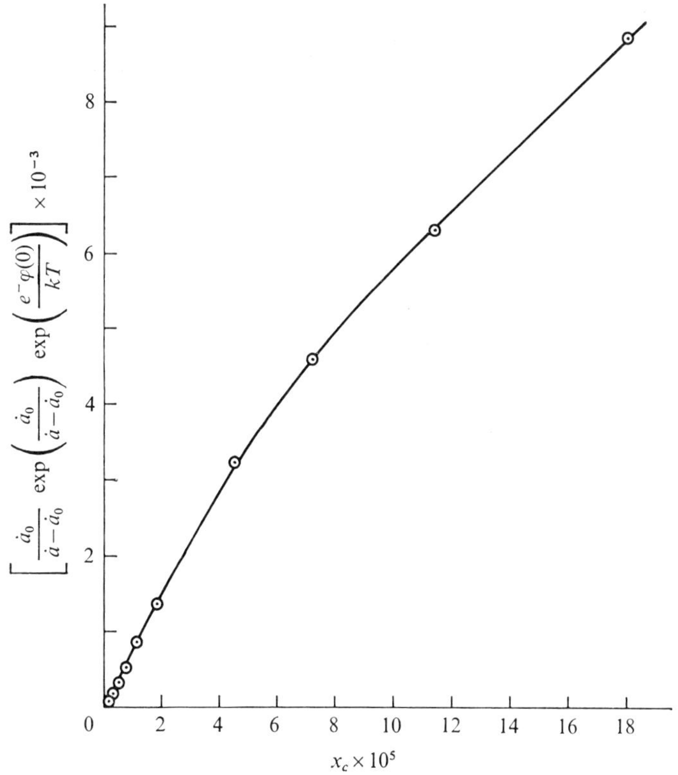

Fig. 3.29. A plot of (3.95) for n-dodecyl trimethylammonium bromide adsorbed at the n-decane–5×10^{-3} M NaCl solution interface at 20 °C. The parameter $\dot{a}_0$ has been taken as 0.25 nm² molecule⁻¹. (Results of Carroll, 1970.)

the desirability of keeping to a minimum interactions between the alkyl chains of the ions, only hydrocarbon–water interfaces are suitable for testing the equations. Figs. 3.28 and 3.29 depict the π versus $1/\dot{a}$ plot and adsorption isotherm respectively for n-dodecyl trimethylammonium bromide at the n-decane–5×10^{-3} M NaCl solution interface. The adsorption results cover a considerably greater range of $\dot{a}$, but are rather less accurate than the data for insoluble monolayers shown in fig. 3.21. The agreement with the theoretical curve in fig. 3.28 is by no means perfect, and cannot be made so by adjustment of the parameter $\dot{a}_0$. Also, the isotherm in fig. 3.29 is not linear, as required by the theory. At higher electrolyte concentrations the deviations from theory become still more serious. By contrast, the agreement between theory and experimental data for un-ionised adsorbed monolayers is excellent, as shown in figs. 3.22 and 3.23. The conclusion that the simple electrical model is inadequate is difficult to avoid.

Appendix

The surface chemical potentials μ^{ads} (in 1.58) and μ_i^σ, as well as the quantity ζ_i have been used in this book. The significance of these three parameters and the reasons for using them are not perhaps immediately obvious and it may be helpful to summarise the situation.

μ^{ads} was used primarily for a system in which a gas is adsorbed on to an inert surface, and was defined as

$$\mu^{\text{ads}} = \left(\frac{\partial A^{\text{ads}}}{\partial N}\right)_{T,V,a}, \tag{1.58}$$

where A^{ads} is the free energy of the adsorbed molecules only, not of the surface as a whole. Expressions derived from μ^{ads} or A^{ads} describe only the properties of the adsorbed molecules (e.g. surface pressure) and not those of the surface (e.g. surface tension). If the tension (γ_0) of the surface is known prior to the adsorption then, of course, it may be possible to predict the tension after adsorption from the expression

$$\pi = \gamma_0 - \gamma.$$

However, for adsorption on to solid surfaces the chief interest is usually in the amount adsorbed, which is directly measured, and the surface pressure. The surface tension in these systems is not measured and may, in any case, not be an equilibrium quantity (§1.2). μ^{ads} is therefore particularly appropriate for the consideration of adsorption at solid surfaces, although it is sometimes used for liquid surfaces (e.g. p. 97).

The terms μ_i^σ and ζ_i are in contrast to μ^{ads}, both defined in terms of the free energy of the surface as a whole, A^σ; i.e.

$$\mu_i^\sigma = \left(\frac{\partial A^\sigma}{\partial n_i^\sigma}\right)_{T,V^\sigma,n_j^\sigma,a}, \tag{1.28}$$

$$\zeta_i = \left(\frac{\partial A^\sigma}{\partial n_i^\sigma}\right)_{T,V^\sigma,n_j^\sigma,\gamma} \tag{1.34}$$

and are therefore particularly appropriate for systems in which A^σ is relatively well-defined, e.g. for liquid as opposed to solid surfaces.

It may be asked why ζ_i is written in the form of (3.30). The reason is that the surface phase is treated in a manner analogous to an ideal bulk solution. From (1.14) the Helmholtz free energy, A^m, of the monolayer is seen to be

$$A^m = -pV^m + \gamma \mathcal{A} + \sum_i \mu_i^m n_i^m$$

which, since $\mathcal{A} = \sum_i n_i^m a_i$, may be written

$$A^m = -pV^m + \sum_i n_i^m(\mu_i^m + \gamma a_i)$$

or, using (1.33),

$$A^m = -pV^m + \sum_i n_i^m \zeta_i.$$

For an ideal solution

$$A = -pV + \sum_i n_i \mu_i^l$$

$$= -pV + \sum_i n_i(\mu_i^{\ominus, l} + RT \ln x_i^l).$$

Correspondingly, ζ_i has been written in (3.30) as

$$\zeta_i = \mu_i^{\ominus, m} + RT \ln x_i^m \qquad (3.30)$$

so that A^m is given by

$$A^m = -pV^m + \sum_i n_i^m(\mu_i^{\ominus, m} + RT \ln x_i^m).$$

The writing of (3.30) is therefore consistent with the assumption that the free energy of the surface phase may be formulated in exactly the same way as that of a bulk phase. This leads, as shown, to (3.33). The term $(\mu_i^{0,m} - \mu_i^{0,l})$ is a standard free energy of adsorption of component *i* from a perfect solution, and is equal to $\gamma_i^0 a_i$.

The reader may find it instructive to consider the difference in physical significance between the standard free energy $(\mu_i^{0,m} - \mu_i^{0,l})$ for adsorption at the surface of a liquid mixture, as given by (3.33), and $\Delta_a \mu^0$ discussed in §1.14 for the adsorption of a gas on an inert surface.

References

Adam, N. K. (1922). *Proc. Roy. Soc.* A **101**, 452.
Adam, N. K. (1968). *Physics and Chemistry of Surfaces* (Dover Publications, New York).
Adam, N. K. & Jessop, G. (1926*a*). *Proc. Roy. Soc.* A **110**, 423.
Adam, N. K. & Jessop, G. (1926*b*). *Proc. Roy. Soc.* A **112**, 362.
Adam, N. K. & Jessop, G. (1926*c*). *Proc. Roy. Soc.* A **112**, 376.
Adam, N. K. & Rosenheim, O. (1929). *Proc. Roy. Soc.* A **126**, 25.
Adamson, A. W. (1967). *Physical Chemistry of Surfaces*, 2nd edition (Interscience): (*a*) pp. 59 and 375; (*b*) chapter 1; (*c*) pp. 87–92.
Aveyard, R. & Briscoe, B. J. (1970). *Trans. Faraday Soc.* **66**, 2911.

Bell, G. M. & Levine, S. (1966). *Z. physik. Chem.* **231**, 289.
Belton, J. W. & Evans, M. G. (1945). *Trans. Faraday Soc.* **41**, 1.
Betts, J. J. & Pethica, B. A. (1960). *Trans. Faraday Soc.* **56**, 1515.
Brooks, J. H. & Pethica, B. A. (1964). *Trans. Faraday Soc.* **60**, 208.
Carroll, B. J. (1970). Ph.D. dissertation, Cambridge.
Clint, J. H., Corkill, J. M., Goodman, J. F. & Tate, J. R. (1968). *J. Colloid & Interface Sci.* **28**, 522.
Davies, J. T. & Rideal, E. K. (1963). *Interfacial Phenomena*, 2nd edition (Academic Press, New York): pp. 251–64.
Defay, R., Prigogine, I., Bellemans, A. & Everett, D. H. (1966). *Surface Tension and Adsorption* (Longmans): (*a*) pp. 271 and 310; (*b*) p. 256; (*c*) chapter 11; (*d*) p. 166; (*e*) p. 77.
Fowkes, F. M. (1963). *J. Phys. Chem.* **67**, 2538.
Fowkes, F. M. (1964). *Ind. Eng. Chem.* **56**, 40.
Frumkin, A. N. (1924). *Z. physik. Chem.* **109**, 34.
Girifalco, L. A. & Good, R. J. (1957). *J. Phys. Chem.* **61**, 904.
Guggenheim, E. A. (1945). *Trans. Faraday Soc.* **41**, 150.
Guggenheim, E. A. (1967). *Thermodynamics*, 5th edition (North Holland): pp. 163–6.
Harkins, W. D. & Brown, F. E. (1919). *J. Amer. Chem. Soc.* **41**, 499.
Hill, T. L. (1960). *Introduction to Statistical Thermodynamics* (Addison-Wesley): pp. 314–18.
Hughes, A. H. & Rideal, E. K. (1933). *Proc. Roy. Soc.* A**140**, 253.
Johnson, R. E. & Dettre, R. H. (1966). *J. Colloid & Interface Sci.* **21**, 610.
Lewis, G. N. & Randall, M. (1961). *Thermodynamics* (revised by K. S. Pitzer & L. Brewer), 2nd edition (McGraw-Hill): pp. 482–4.
McBain, J. W. & Humphreys, C. W. (1932). *J. Phys. Chem.* **36**, 300.
McBain, J. W. & Swain, R. C. (1936). *Proc. Roy. Soc.* A**154**, 608.
Mitchell, R. W. (1969). Ph.D. dissertation, University of Hull.
Moilliet, J. L., Collie, B. & Black, W. (1961). *Surface Activity* (E. & F. N. Spon, London): pp. 64–73.
Ono, S. & Kondo, S. (1960). *Handbuch der Physik* (ed. Flügge), (Springer, Berlin): vol. 10, p. 134.
Onsager, L. & Samaras, N. N. T. (1934). *J. Chem. Phys.* **2**, 528.
Prigogine, I. & Maréchal, J. (1952). *J. Colloid Sci.* **7**, 122.
Randles, J. E. B. (1963). *Advances in Electrochemistry and Electrochemical Engineering* (ed. P. Delahay and C. W. Tobias), (Interscience): vol. 3, p. 1.
Schmutzer, E. (1955). *Z. phys. Chem.* (*Leipzig*), **204**, 131.
Shafrin, E. G. & Zisman, W. A. (1967). *J. Phys. Chem.* **71**, 1309.
Smith, T. (1967). *J. Colloid Sci.* **23**, 27.

4 Polarised and non-polarised electrode surfaces

4.1. Introduction. Electrical double layers have so far been discussed only for air– and hydrocarbon–water interfaces. It has been shown that some of the properties of these interfaces can be explained qualitatively or semi-quantitatively by the use of the theory given in chapters 1 and 2. For the purpose of testing the electrical double layer theories, however, air– and hydrocarbon–water interfaces are not the most suitable systems. Although ion adsorption can be measured by means of the application of the Gibbs equation to interfacial tension data, the only potential that is measurable is the change in Volta potential and, as shown in chapter 2, this involves both dipole and ionic double layer terms, both of which are unknown. In this chapter, two more suitable types of system will be described, one of which is said to have an *ideally polarised* interface and the other, an *ideally non-polarised* interface. Some discussion of these terms may be helpful. When a conductor is in contact with an aqueous solution, the application of a potential to either phase will produce a flow of charge across the interface. The rate of flow of the charge is, however, determined by the activation energies required for the ions or electrons to cross the interface and may vary enormously from one system to another. It is convenient to consider the two extreme possibilities which arise either when the activation energy is so high that the rate of flow of charge is effectively zero or, when the activation energy is so small that there is effectively no barrier to the flow of charge.

In the first instance the surface is said to be ideally polarised. On changing the applied potential, charge is accumulated or lost from the phase boundaries on either side of the interface. Although no charge crosses the interface an equilibrium condition is established, both between the two surface and bulk phases and also across the interface. The latter equilibrium is, however, electrostatic and mechanical rather than electrochemical.

In the second instance the surface is said to be ideally non-polarised. Ions or electrons pass immediately and freely across the interface and continue to do so for as long as a potential other than an equilibrium value exists. The system constitutes a reversible electrode which is in electrochemical equilibrium with the aqueous solution.

No interfaces are ideal in either of the senses discussed above. Under certain circumstances, however, some metal–aqueous solution interfaces are almost ideally polarised, and some crystalline solids in contact with solutions of their own ions are almost ideally non-polarised. Mercury in contact with aqueous solutions is a good example of the first type and solid silver iodide in a solution containing silver and iodide ions, a good example of the second.

Mercury systems are extremely convenient as, being entirely liquid, the interfacial tension may be measured. For this reason, these systems have provided much information concerning electrical double layers. The silver iodide type of system, while of lesser value than mercury as a means of testing double layer theory, has other useful features. For example, silver iodide readily forms suspensions in aqueous solutions and, in consequence, has been more valuable than mercury in the study of colloid stability.

4.2. The mercury–electrolyte solution interface: thermodynamic theory of electrocapillarity. Mercury has a high overvoltage and, in contact with aqueous solutions, it behaves as an ideal polarised electrode over a wide range of applied potential; i.e. when the potential difference between the phases is altered within this range, equilibrium is re-established without the transfer of any charge across the interface. It is therefore possible in this system to vary the potential difference between the phases without varying the composition of the solution, and also to vary the composition of the solution at constant potential difference. In order to describe the system it is necessary to extend the Gibbs equation (1.40) to cover the possibility that the electrical potential difference between the phases may be an independent variable. Various treatments of this problem have been given, of which the most elegant and general is probably that of Grahame & Whitney (1942). These authors derive a thermodynamic relationship for a system containing either a polarised or a non-polarised metal–solution interface in which there may be any number of ionic or non-ionic components. The full arguments behind this approach are, however, quite complicated and are probably unnecessary for a first acquaintanceship with the subject. The treatment given here is analogous to that of Parsons (1954). While this approach lacks generality it is nevertheless rigorous and easily understood. Although a very simple system will be dis-

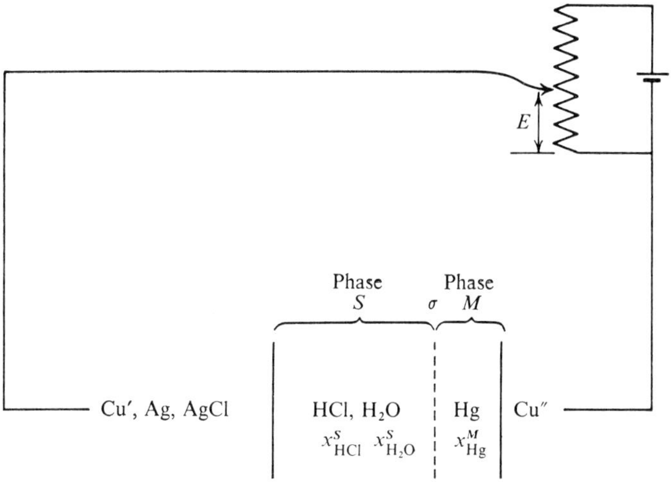

Fig. 4.1. A circuit for the study of the mercury–aqueous solution interface.

cussed, more complex systems can readily be treated by the same methods.

The cell and circuit to be considered are illustrated in fig. 4.1. The phases and mole fractions of the components are as indicated. The mole fraction of the mercury has been included as a formality. In the present system this will be constant and equal to unity but, in general, there may be other components in the mercury phase (see e.g. Parsons' article). The interface is considered as a region of finite thickness and will be treated as a surface phase (§1.6). As a consequence of the choice of HCl as the electrolyte, the presence of hydroxyl ions in the system will be ignored.

A convenient starting point is the Gibbs equation (1.40) in which constant temperature has been assumed. Electrochemical potentials are written in place of chemical potentials, so that

$$-\mathrm{d}\gamma = \sum_i \Gamma_i \mathrm{d}\tilde{\mu}_i. \tag{4.1}$$

Here, Γ_i is the number of moles of species *i* per unit area of the interfacial phase but, for clarity in this treatment, the superscript *s* (§1.8) is omitted. The electrochemical potential is written as described in §2.2 in terms of its 'chemical' component, μ_i, the inner potential of the phase, φ, and the valence (including the sign) z_i, of the component, i.e.

$$\tilde{\mu}_i = \mu_i + z_i e^- \varphi. \tag{4.2}$$

For an uncharged component $z_i = 0$, and the electrochemical potential is identical to the chemical potential.

Equation (4.1) is for an electrically neutral system at equilibrium and the summation over the components must include all the charged particles present in both bulk phases and the interfacial phase. For the system depicted in fig. 4.1 these components are conveniently chosen as H^+ and Cl^- in the aqueous phase, and Hg^+ and electrons (e) in the mercury phase. The expansion of (4.1) then reads

$$-d\gamma = \Gamma_{Hg^+}\, d\tilde{\mu}^M_{Hg^+} + \Gamma_e\, d\tilde{\mu}^M_e + \Gamma_{H^+}\, d\tilde{\mu}^S_{H^+} + \Gamma_{Cl^-}\, d\tilde{\mu}^S_{Cl^-} + \Gamma_{H_2O}\, d\mu^S_{H_2O}. \quad (4.3)$$

The equilibria in the two bulk phases give the following relationships. In the metal phase,

$$\mu^M_{Hg} = \tilde{\mu}^M_{Hg^+} + \tilde{\mu}^M_e, \quad (4.4)$$

and for the electrons in the mercury and the copper wire,

$$\tilde{\mu}^M_e = \tilde{\mu}^{Cu''}_e. \quad (4.5)$$

Therefore, from (4.4) and (4.5),

$$\Gamma_{Hg^+}\, d\tilde{\mu}^M_{Hg^+} + \Gamma_e\, d\tilde{\mu}^M_e = \Gamma_{Hg^+}\, d\mu^M_{Hg} - (\sigma^M/e^-)\, d\tilde{\mu}^{Cu''}_e \quad (4.6)$$

where

$$\sigma^M = (\Gamma_{Hg^+} - \Gamma_e)\, e^-. \quad (4.7)$$

The quantity σ^M is thus the contribution to the interfacial charge of the components of the metal phase. In the aqueous phase,

$$\mu^S_{HCl} = \tilde{\mu}^S_{H^+} + \tilde{\mu}^S_{Cl^-} \quad (4.8)$$

and therefore,

$$\Gamma_{H^+}\, d\tilde{\mu}^S_{H^+} + \Gamma_{Cl^-}\, d\tilde{\mu}^S_{Cl^-} = \Gamma_{H^+}\, d\mu^S_{HCl} - (\sigma^S/e^-)\, d\tilde{\mu}^S_{Cl^-} \quad (4.9)$$

where

$$\sigma^S = (\Gamma_{H^+} - \Gamma_{Cl^-})\, e^-. \quad (4.10)$$

Thus σ^S is the contribution to the interfacial charge of the components of the aqueous phase.

As mentioned above, the whole system must be electrically neutral. The two bulk phases must also be electrically neutral, and therefore the interfacial phase must be neutral. It follows that

$$\sigma^M + \sigma^S = 0. \quad (4.11)$$

As the interface is ideally polarised, no charged component will be found in both phases. The two contributions to the interfacial charge σ^M and σ^S are therefore segregated in a direction normal to the interface and each can be regarded as numerically equal to the interfacial charge density. From (4.3), (4.6), (4.9) and (4.11),

$$-d\gamma = \Gamma_{Hg^+} d\mu_{Hg}^M + \Gamma_{H^+} d\mu_{HCl}^S - (\sigma^M/e^-)[d\tilde{\mu}_e^{Cu''} - d\tilde{\mu}_{Cl^-}^S] + \Gamma_{H_2O} d\mu_{H_2O}^S. \qquad (4.12)$$

At the silver/silver chloride electrode it is easily deduced that

$$d\tilde{\mu}_{Cl^-}^S = d\tilde{\mu}_e^{Cu'}. \qquad (4.13)$$

Furthermore, the copper wires Cu' and Cu" have the same composition and therefore, as argued in §2.2 and §2.15,

$$d\tilde{\mu}_e^{Cu'} - d\tilde{\mu}_e^{Cu''} = d(\varphi^{Cu''} - \varphi^{Cu'}) = dE. \qquad (4.14)$$

Substitution of (4.13) and (4.14) into (4.12) gives

$$-d\gamma = \sigma^M dE + \Gamma_{Hg^+} d\mu_{Hg}^M + \Gamma_{H^+} d\mu_{HCl}^S + \Gamma_{H_2O} d\mu_{H_2O}^S. \qquad (4.15)$$

For each phase a form of the Gibbs–Duhem equation may be written. In the present example the metal phase consists of only one electrically neutral component (Hg) and therefore $d\mu_{Hg}^M = 0$. For the aqueous phase

$$x_{HCl}^S d\mu_{HCl}^S + x_{H_2O}^S d\mu_{H_2O}^S = 0. \qquad (4.16)$$

If, from these considerations, $d\mu_{H_2O}^S$ is eliminated from (4.15) the result is

$$-d\gamma = \sigma^M dE + \left(\Gamma_{H^+} - \frac{x_{HCl}^S}{x_{H_2O}^S}\Gamma_{H_2O}\right) d\mu_{HCl}^S, \qquad \blacktriangleleft(4.17)$$

where the term in brackets is the surface excess of H^+.

The expression (4.17) is the general electrocapillary equation for the system depicted in fig. 4.1. In practice the equation is used under conditions either of constant composition or of constant potential difference. For the former case, (4.17) reduces to

$$\left(\frac{\partial \gamma}{\partial E}\right)_{T, p, \mu^M, \mu^S} = -\sigma^M = \sigma^S. \qquad (4.18)$$

This equation is usually referred to as the Lippmann equation. It shows that for an ideally polarised electrode the interfacial charge density may be obtained by taking the slope of the plot of interfacial tension against applied potential.

The separation of charge in a direction normal to the interface gives the interface the properties of a capacitor. The interrelationship of charge and potential for this capacitor is not, in general, linear and in consequence it is necessary to recognise two types of capacitance, differential and integral. The differential capacitance is given by

$$C' = \frac{\partial \sigma^M}{\partial E} = -\frac{\partial^2 \gamma}{\partial E^2} \tag{4.19}$$

and the integral capacitance by

$$C = \frac{\sigma^M}{E - E^{\text{ecm}}}. \tag{4.20}$$

E^{ecm} is the potential at the electrocapillary maximum, where the charge on the interface is zero. The phenomenon of the electrocapillary maximum is described in more detail in §4.3. The differential capacitance can be obtained from the plot of γ versus E. The integral capacitance can also be obtained from this plot but is, in addition, directly measurable, as mentioned below.

If, instead of the composition, the applied potential is kept constant, (4.17) becomes

$$-\left(\frac{\partial \gamma}{\partial \mu^S_{\text{HCl}}}\right)_{T, p, E, \mu^M} = \Gamma_{\text{H}^+} - \frac{x^S_{\text{HCl}}}{x^S_{\text{H}_2\text{O}}} \Gamma_{\text{H}_2\text{O}} \tag{4.21}$$

and the surface excess of H^+ can be obtained by the usual method of plotting γ versus log $\boldsymbol{a}$. In more complex systems the surface excesses of all ionic and non-ionic components (with the exception of the reference components) can be found in a similar way.

4.3. The mercury–electrolyte solution interface: experimental methods and results. The measurement of interfacial tension and of applied potential is usually carried out in a capillary electrometer similar to that devised by Lippmann. Such an instrument is illustrated schematically in fig. 4.2. The measurement of the applied potential E or of changes in E is self-evident provided an appropriate reference electrode is chosen. The interfacial tension is measured by the capillary rise method. Owing to the large contact angle between mercury and glass in the presence of water (*ca.* 180°) a pressure (given by the height, h) has to be applied to the mercury in order to force it down to a given point in the fine, slightly tapered capillary at the bottom of the column. The interfacial tension is calculated from the value of h. It is common to calibrate a capillary with a system of known properties rather than to use the apparatus for absolute tension measurement.

The type of results obtained is illustrated in fig. 4.3 by the electrocapillary curves for various uni-univalent electrolytes. At zero applied

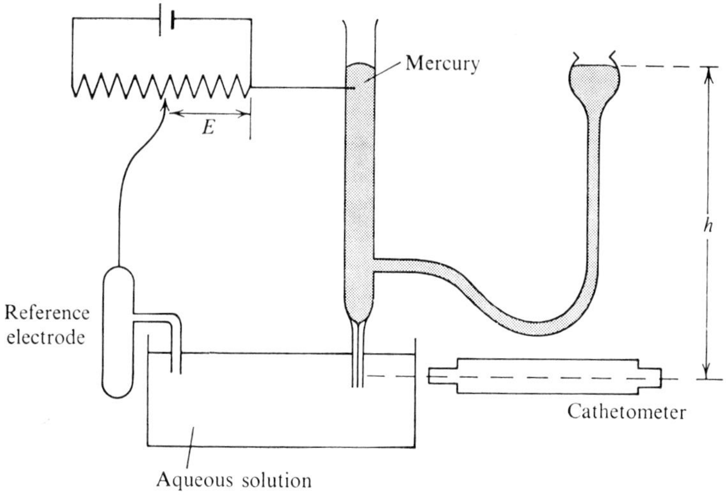

Fig. 4.2. A Lippmann capillary electrometer (see text).

potential the interfacial tension is lowered by the adsorption of the electrolyte ions, the magnitude of the lowering and of the adsorption being a function of the nature of the anion. As the applied potential increases in the negative sense the tensions pass through a maximum (the electrocapillary maximum) where, according to (4.18), the net charge on the interface is zero. At still more negative potentials the tension falls but eventually in a way independent of the nature of the anion. For a system containing different cations and a common anion the picture is reversed. However, for inorganic cations the curves are almost coincident even at strongly negative potentials. It is mainly for organic cations, e.g. trimethyl- or triethylammonium ions, that large differences appear.

The interpretation of the curves at a qualitative level is fairly obvious. At high negative applied potentials $\partial\gamma/\partial E$ and hence, by (4.18), σ^S are positive; i.e. cations are attracted to the surface from the aqueous solution and lower the tension. The anions are repelled. The fact that different inorganic cations give similar results suggests that there are no specific forces of attraction and that only coulombic forces are involved. At the potential of the electrocapillary maximum the charge on the interface is zero and the adsorption of cations and anions is equal. At relatively low negative potentials the anions are adsorbed preferentially and the tension falls. However, the markedly different results for

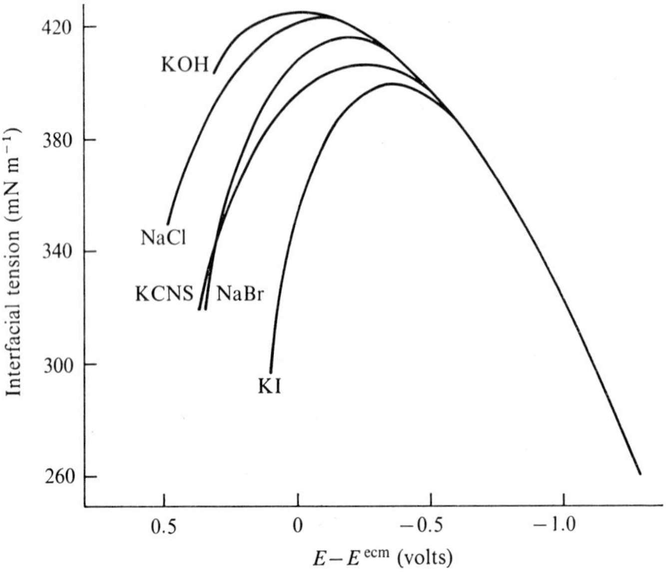

Fig. 4.3. The interfacial tension between mercury and some aqueous solutions of uni-univalent electrolytes (redrawn from Grahame, 1947).

different anions suggests that, in addition to coulombic forces, there must be short range forces of interaction between the anions and the mercury surface which are highly specific to the type of anion. As can be seen from fig. 4.3 the more polarisable anions are the more strongly adsorbed, although it should not be assumed that polarisability is the only factor involved.

The validity of this type of explanation, and indeed the quantitative confirmation of the validity of the Lippmann equation, have been well established by experiment. Thus, the conclusion that the sign of the charge on one side of the interface changes on passing through the electrocapillary maximum may be confirmed by forcing the mercury to drop from a capillary to the bottom of the aqueous phase (fig. 4.4). As the droplets of mercury form at the tip of the capillary, charge is attracted to or repelled from the expanding interface in order to maintain the equilibrium value of the charge density (σ^M or σ^S). In consequence, a current flows in the external circuit connecting the mercury column to the pool in the aqueous phase. The direction of the current flow then indicates the sign of the surface charge. With the biasing circuit the surface charge may be examined for any chosen value of the

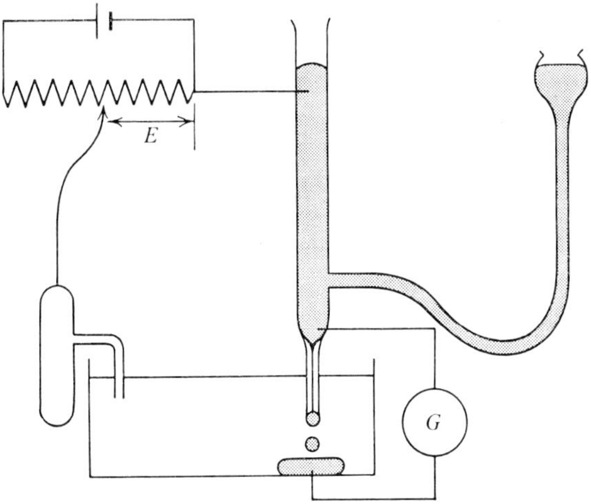

Fig. 4.4. The determination of the charge at the mercury–aqueous solution interface. Both the sign and the magnitude of the charge may be estimated from the response of the galvanometer (G) and the number and surface area of the drops.

applied potential. The magnitude of the surface charge may also be determined by this type of experiment. The experimental results fully confirm the predictions of the Lippmann equation.

The electrical capacitance as determined from the Lippmann equation has also been satisfactorily confirmed by direct measurements of the capacitance using a.c. bridge techniques (Grahame, 1947).

Although the system which has been most studied and which is primarily discussed here is that of mercury against aqueous solutions, electrocapillary investigations of other metals have also been carried out. Where these metals are solid the interfacial tensions have been deduced from the contact angles of bubbles resting against their surfaces. As may be appreciated, however, there are usually difficulties in the deduction of solid–liquid tensions by this method and at best the electrocapillary results are probably not as accurate as those for mercury.

4.4. Tests of electrical double layer theory. From what has been said above it will have become obvious that the mercury–aqueous solution system is extraordinarily well suited to testing the theories of the electrical double layer described in chapter 2. In the mercury systems it is possible to measure surface charge density (4.18), ion adsorption

(4.21) and double layer capacitance (4.19). It can also be argued that, owing to the infinite or, at least, very high dielectric constant of the mercury, the charge on the metal will be effectively at the phase boundary. It follows that the whole potential change at the interface should occur in the aqueous phase. This is precisely the situation considered in the Gouy–Chapman and Stern theories.

For the purpose of testing the Gouy–Chapman theory and the ionic size aspect of the Stern theory, it is desirable to examine a system in which specific adsorption of ions at the mercury surface is absent. As mentioned in connection with fig. 4.3, specific adsorption of the anion, at least, appears to occur in most electrolytes. For sodium fluoride, however, various pieces of evidence suggest that specific adsorption of both the anion and the cation are minimal and possibly negligible. Grahame (1954, 1957) has studied the differential capacitance for mercury against sodium fluoride solutions in considerable detail. The differential capacitance is predicted by the Gouy–Chapman theory, and may be obtained by differentiation of (2.31). Thus

$$\sigma = \sigma^M = (8N(\infty)\epsilon_r\epsilon_0 kT)^{\frac{1}{2}} \sinh\frac{ze^-}{2kT}(\varphi(0)-\varphi(\infty)) \tag{2.31}$$

and

$$C' = \frac{\partial\sigma^M}{\partial(\varphi(0)-\varphi(\infty))} = \left(\frac{2z^2(e^-)^2N(\infty)\epsilon_r\epsilon_0}{kT}\right)^{\frac{1}{2}} \cosh\frac{ze^-}{2kT}(\varphi(0)-\varphi(\infty)). \tag{4.22}$$

As σ^M is known from experiment via the Lippmann equation (4.18), first $(\varphi(0)-\varphi(\infty))$ and then C' may be calculated. Theory and experiment are compared in fig. 4.5. At potentials close to that corresponding to the electrocapillary maximum ($\sigma^M \approx 0$) and for 0.001 M NaF the agreement is fair, but at other potentials in 0.001 M, and at all potentials for higher concentrations the agreement is poor.

One obvious possible reason for the discrepancies lies in the assumption inherent in the Gouy–Chapman theory that the whole double layer potential drop exists across the diffuse layer, or, in other words, that $\varphi(0)$ may be identified with φ^M, the potential at the mercury surface. The theory of Stern (§2.7) is intended to correct for such a discrepancy. For the model depicted in fig. 2.7 the potentials may be written

$$\varphi(0)-\varphi(\infty) = (\varphi(0)-\varphi(\delta))+(\varphi(\delta)-\varphi(\infty)) \tag{4.23}$$

and

$$\frac{\partial(\varphi(0)-\varphi(\infty))}{\partial\sigma^M} = \frac{\partial(\varphi(0)-\varphi(\delta))}{\partial\sigma^M} + \frac{\partial(\varphi(\delta)-\varphi(\infty))}{\partial\sigma^M}. \tag{4.24}$$

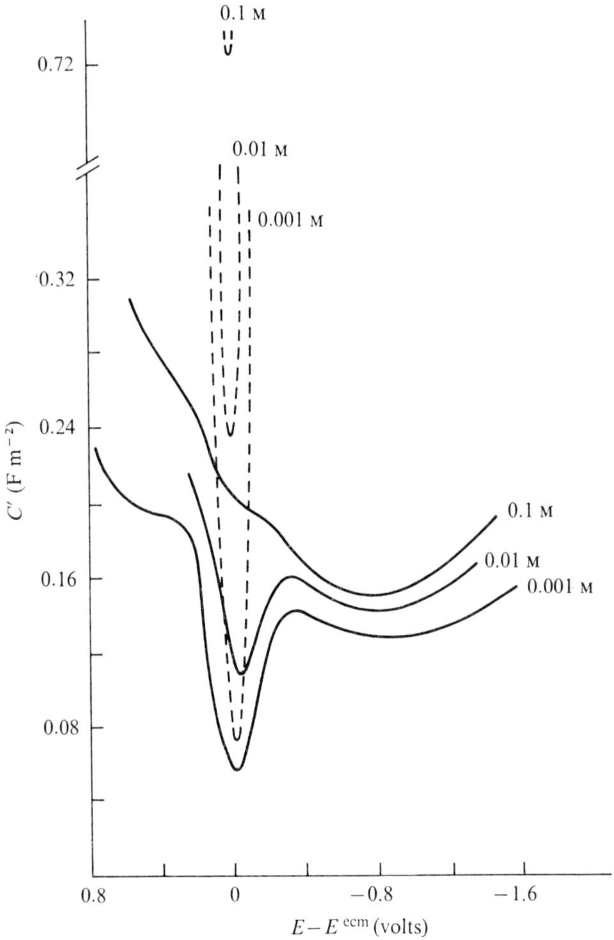

Fig. 4.5. The differential capacitance of the mercury electrode in sodium fluoride solutions. The full curves are the experimental results of Grahame (1947). The dashed curves are calculated from (2.31) and (4.22).

The various terms of (4.24) are the reciprocals of the differential capacitances of the respective regions of the double layer and thus,

$$\frac{1}{C'} = \frac{1}{C'_{st}} + \frac{1}{C'_d}. \tag{4.25}$$

According to the Stern model the $\varphi(0)$ of the Gouy theory should be identified with $\varphi(\delta)$. Consequently the C'_d of (4.25) is equal to the C' of

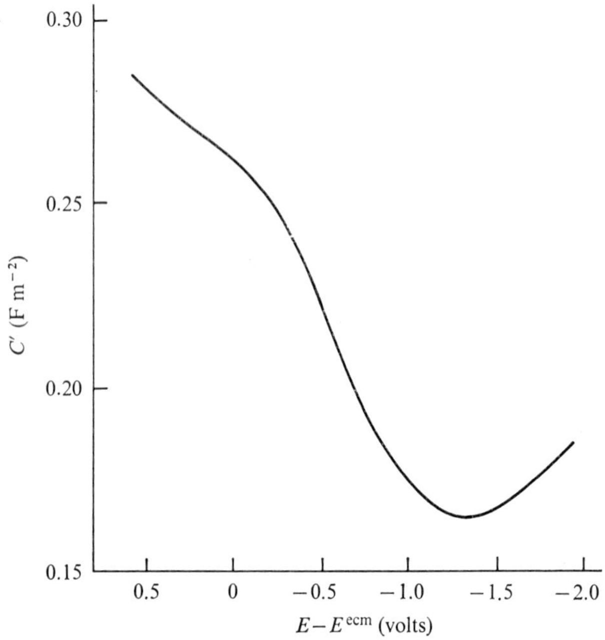

Fig. 4.6. The differential capacitance (C'_{st}) of the non-diffuse part of the double layer calculated from data for 1.0 M sodium fluoride. (After Grahame, 1947.)

(4.22) and can therefore be evaluated as previously. If, then, the capacitance of the Stern layer or molecular capacitor can be found, C' can be calculated and compared with the experimental data. In order to obtain C'_{st}, Grahame (1947) made use of the fact that in concentrated electrolyte solutions the differential capacitance of the diffuse double layer becomes so high that the last term in (4.25) is almost negligible, and $C' \simeq C'_{st}$. His results for 1.0 M NaF are shown in fig. 4.6. The values of C' for lower concentrations of NaF (based on C'_{st} from fig. 4.6 and C'_{d} from (4.23) and (4.24)) are compared with the measured values in figs. 4.7–4.9. Grahame's conclusion from these data is that while the agreement is not perfect it indicates that the Gouy–Chapman diffuse double layer theory 'is accurate enough to be useful'. The Stern ionic size correction as exemplified by (4.25) also appears to be on the right lines, in that it evidently contributes to the above agreement with experimental data. However, the marked variation of the capacitance of the Stern layer with applied potential indicated in fig. 4.6 requires comment.

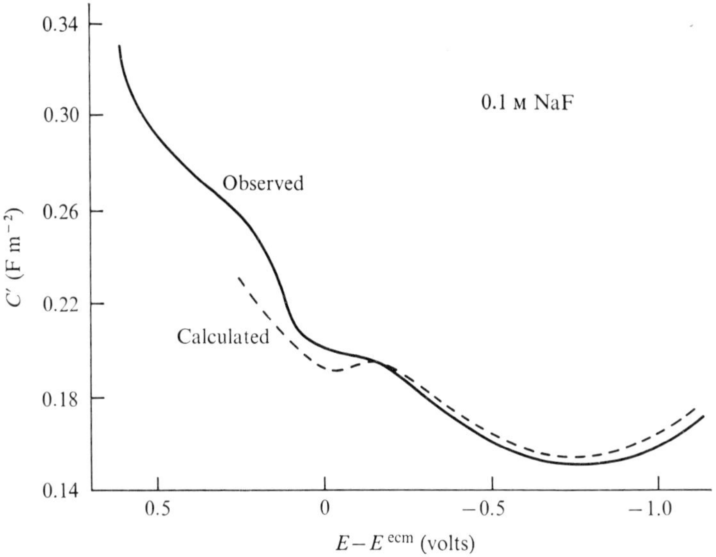

Fig. 4.7. The calculated and observed differential capacitance of mercury in contact with 0.1 M sodium fluoride. (After Grahame, 1947.)

The Stern layer capacitance depends on, among other factors, the distance of closest approach of the ions of the ionic double layer to the mercury surface. On the positive side of the electrocapillary maximum the most significant ion will be the anion, while on the negative side it will be the cation. If the anion and cation have different effective radii then it is to be expected that the Stern layer capacitance will differ from one side of the electrocapillary maximum to the other, and, indeed, should show a continuous variation with applied potential. While this certainly happens, the results are by no means simple and the theoretical treatment of this phenomenon, which involves also the question of the state of the water in the Stern layer, is beyond the scope of this chapter.

In the above discussion only results for sodium fluoride have been mentioned, where it has been assumed that for moderate polarising potentials no specific adsorption of ions occurs. In general, however, specific adsorption of anions does occur. It is usually assumed that the specifically adsorbed ions lose their water of hydration on that side of the ion closest to the interface and thereby are able to approach more closely to the mercury. It has, in fact, been suggested that these ions form covalent bonds with the mercury. Thus, in place of the single

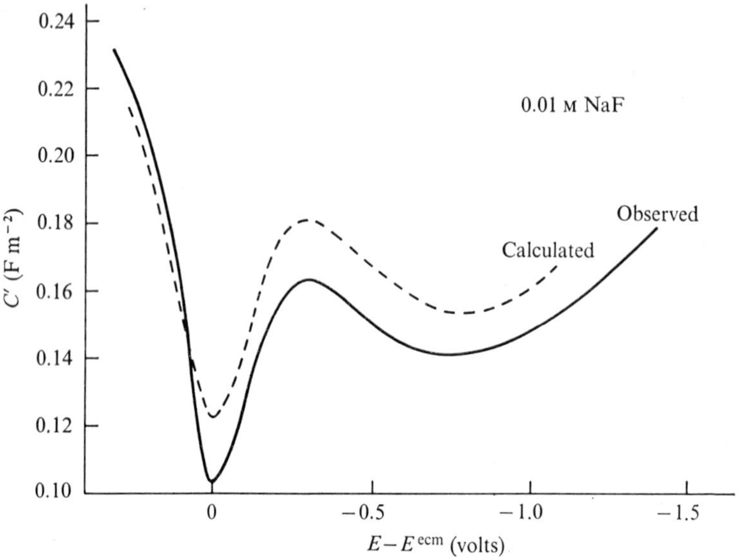

Fig. 4.8. The calculated and observed differential capacitance of mercury in contact with 0.01 M sodium fluoride. (After Grahame, 1947.)

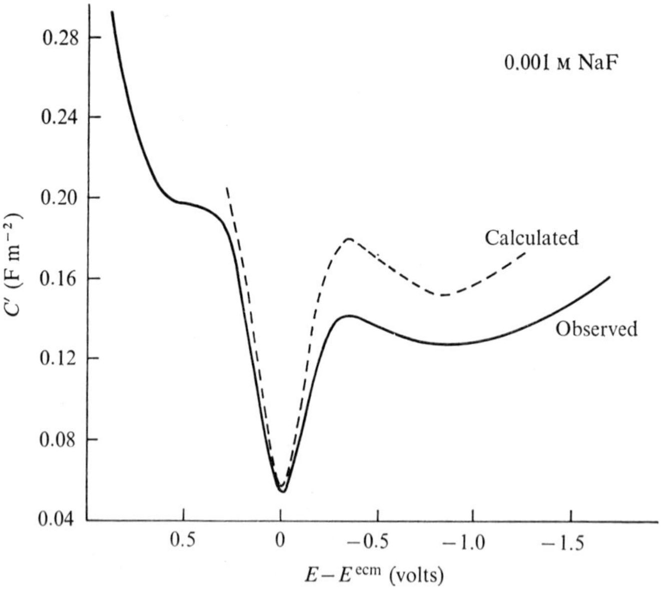

Fig. 4.9. The calculated and observed differential capacitance of mercury in contact with 0.001 M sodium fluoride. (After Grahame, 1947.)

plane of ion centres on which the original Stern theory was based (fig. 2.7), it is necessary to postulate two planes, the outer, to correspond with the plane at distance δ invoked previously for the hydrated non-specifically adsorbed ions, and the inner, for the specifically adsorbed ions. The two planes are known as the *outer* and *inner Helmholtz planes*. The quantitative treatment of this model has been attempted by several authors but the problem is complicated and will not be discussed here. The diagrams given by Grahame (1947) and reproduced in fig. 4.10 perhaps illustrate most simply the probable variation of potential in the double layer when specific adsorption is present. Only average potentials are depicted, i.e. discrete charge effects are ignored, and no attempt has been made to represent potential changes arising from oriented water dipoles. Fig. 4.10(*a*) shows the situation at the electrocapillary maximum. The field within the mercury is, as always, zero. As there is no net charge on the mercury surface, electrostatic theory requires that the field on the aqueous side of the boundary should also be zero. At the inner Helmholtz plane some anions will be specifically adsorbed (negative signs in small circles) and, in consequence, this plane will carry a negative charge density. This negative charge must be balanced by an equal and opposite net charge consisting predominantly of cations situated at the outer Helmholtz plane and in the diffuse layer. The potential is therefore assumed to rise linearly between the inner and outer Helmholtz planes and, beyond the latter, to show the usual diffuse layer variation into the bulk aqueous phase. The ions with dashed outlines, or 'ghosts' as Grahame calls them, represent the ions that would have been present if the mercury surface were absent.

Fig. 4.10(*b*) depicts the situation for negative polarisation of the mercury surface. (The excess of electrons is shown in the form of negative signs on the metal side of the interface.) Anions are repelled from the surface and no specific adsorption occurs; the inner Helmholtz layer is therefore empty. The negative electronic charge on the mercury is balanced by a net positive ionic charge in the outer Helmholtz layer and in the diffuse layer, and the variation of potential is as originally discussed in connection with the Stern theory in §2.7.

In Fig. 10(*c*) the mercury surface is assumed to be positively polarised. (The positive signs indicate a surface deficit of electrons.) Specific adsorption of anions is extensive and more than compensates for the positive charge on the mercury. In consequence there is a steep fall in potential from the mercury surface to the inner Helmholtz plane. From this plane outwards there is a net negative charge to be balanced and,

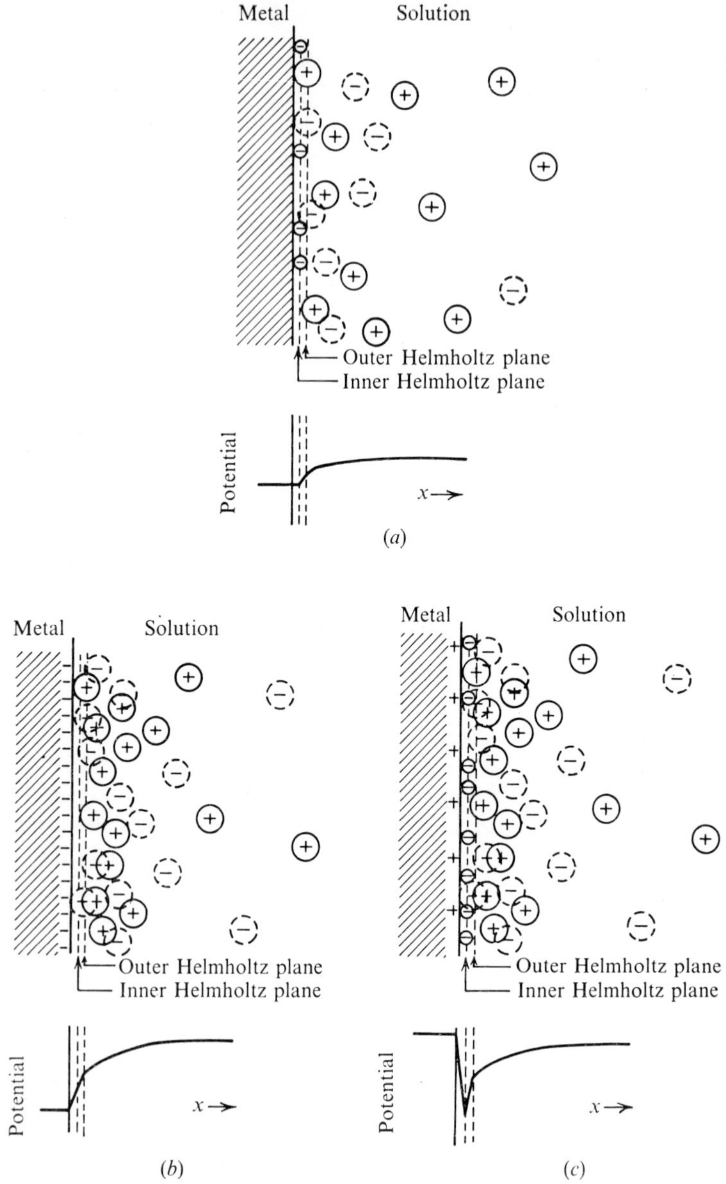

Fig. 4.10. Schematic representations of the electrical double layer at a mercury–aqueous solution interface under different conditions of polarisation. (After Grahame, 1947.)

(*a*) At the potential corresponding to the electrocapillary maximum. Small circles represent specifically adsorbed ions. Large circles represent solvated ions. Dotted circles represent 'ghosts', or ions which would be present if the double layer were not there.

(*b*) With the mercury negative relative to the aqueous phase. Specifically adsorbed ions are absent and positive ions are concentrated near the surface more than in (*a*). The concentration of 'ghosts' is also increased.

(*c*) With the mercury positive relative to the aqueous phase. Specific adsorption of anions is extensive. The diffuse layer is identical with that depicted in (*b*).

as before, this is achieved by a net positive ionic charge in the outer Helmholtz and diffuse layers.

To conclude the discussion of mercury–aqueous solution systems it is appropriate to return to the point made at the outset, that the advantage of these systems for the study of electrical double layers lies in the peculiarity that they contain a liquid metal. This does not necessarily, however, detract from their value in propagating the understanding of other systems. Grahame (1947) writes:

> At the surface of a non-metallic particle in a salt solution there exists an electrical double layer no different in principle from those which we have discussed above. It is convenient to regard the neutral surface of the particle as the counterpart of the metallic surface. Adsorbed ions then produce a charge on the particle, but since this is not an electronic charge, the particle is analogous to the metallic surface *at its electrocapillary maximum.* Beyond this layer of adsorbed ions lies the diffuse layer, whose maximum potential φ (outer Helmholtz plane) is...

There follows a brief comment on the possible identity of the potential of the outer Helmholtz plane and the electrokinetic (zeta) potential. This question has already been discussed in §2.9 where the maximum potential of the diffuse layer has been denoted as $\varphi(\delta)$.

4.5. The silver iodide–electrolyte solution interface: the experimental system and underlying theory. The quantitative treatment of silver iodide solution interfaces poses more problems than does the mercury system as will become evident from the following paragraphs. The theory starts from the experimental observation that the interface between silver iodide and a solution of its own ions is not polarisable (at least, for small current densities) and constitutes a reversible half cell. A suitable experimental arrangement may be obtained by combining this half cell with, for example, a calomel electrode:

Pt, Hg, Hg_2Cl_2	KCl (sat.)	Silver iodide suspension in solution containing Ag^+ and I^- ions	AgI, Ag, Pt.

The presence of the silver iodide suspension is to facilitate the surface charge determination, which will be mentioned later. The potassium chloride must be connected to the silver iodide solution by means of an appropriate salt bridge in order to avoid the removal of silver ions through silver chloride formation.

The e.m.f. (E) of the cell is made up of contributions from the calomel electrode, such diffusion potentials as may be associated with the salt bridge, and the silver iodide electrode. Thus

$$E = E(\text{calomel}+\text{diffusion})+E^0_{\text{Ag}}+\frac{RT}{F}\ln \boldsymbol{a}_{\text{Ag}^+}$$

$$= E(\text{calomel}+\text{diffusion})+E^0_{\text{I}}-\frac{RT}{F}\ln \boldsymbol{a}_{\text{I}^-}. \qquad (4.26)$$

If the diffusion potential is effectively constant, changes in E corresponding to changes in the composition of the solution may be attributed entirely to the silver iodide electrode. Furthermore the changes in E can be equated to the changes in the Galvani potential difference between the aqueous and silver iodide phases. In differential form, (4.26) then becomes

$$\mathrm{d}E = \mathrm{d}(\Delta\varphi) = \frac{RT}{F}\mathrm{d}\ln \boldsymbol{a}_{\text{Ag}^+}$$

$$= -\frac{RT}{F}\mathrm{d}\ln \boldsymbol{a}_{\text{I}^-}. \qquad (4.27)$$

Studies of the surface chemistry of silver iodide are based principally on the examination of the above cell. The procedure usually adopted is, in outline, as follows. Firstly, the concentrations of the potential-determining ions Ag^+ and I^- which correspond to zero net charge on the silver iodide surface are established. Secondly, the surface charge densities on the silver iodide are varied by varying the concentrations of Ag^+ and I^-. The actual values of the charge densities are deduced from a knowledge of the surface excesses of Ag^+ and I^- relative to those for surfaces of zero net charge. Thirdly, the cell e.m.f. for any arbitrary concentration of Ag^+ and I^- relative to that for zero charge concentrations is interpreted in terms of double layer potentials. It is then possible to calculate (from the slope of the charge versus potential curves) values of the electrical double layer capacitance and to attempt an analysis of the results in a rather similar way to that for mercury systems.

Several aspects of this procedure involve complex problems and each stage will be described in more detail.

The zero point of charge may be determined in the following way from the measurement of electrokinetic mobilities or potentials (§2.9). If no ions other than Ag^+ or I^- are specifically adsorbed on to the silver iodide surface the ionic double layer charge should be entirely in the diffuse region and, in consequence, when the electrokinetic potential is zero the net surface charge should also be zero. In practice the determination of the zero point of charge is usually carried out in a dilute solution of an 'inert' electrolyte, such as 10^{-3} M KNO_3. The concentrations of Ag^+ or I^- are adjusted by the addition of small amounts of $AgNO_3$ or KI until the electrokinetic potential is zero. The final concentrations of Ag^+ and I^- are usually much smaller than that of the inert electrolyte and the activity coefficients of the Ag^+ and I^- may be assumed to be constant. In consequence, after calibration of the cell with known concentrations of Ag^+ and I^-, the concentrations of these species in any unknown system may be deduced from one e.m.f. measurement. It is found that for zero surface charge the equilibrium concentrations of Ag^+ and I^- are in the region of 4×10^{-6} M and 3×10^{-11} M respectively. In general the zero point of charge depends on the composition of the aqueous phase, but there are electrolytes such as KNO_3 which have a negligible effect even at fairly high concentrations. Other methods for the determination of the zero point of charge have been described by Overbeek (1952). For carefully conducted experiments the results are found to be in reasonable agreement.

The determination of the surface charge density of the silver iodide is carried out in the cell as illustrated above. The Ag^+ and I^- concentrations are adjusted until the e.m.f. corresponding to the zero point of charge is attained. A known quantity of Ag^+ or I^- is then added and, after equilibrium the e.m.f. is again recorded. The adsorption of Ag^+ or I^- by the suspension and electrode surfaces can then be calculated from the initial and final concentrations (deduced from the e.m.f. measurements) together with the knowledge of how much Ag^+ or I^- was added to the system. (The reason for having a suspension of silver iodide in the cell is now evident. It is necessary to have sufficient silver iodide surface available for adsorption, to make the differences between the amounts of Ag^+ or I^- in the initial and final suspensions accurately measurable.) Provided all the adsorbed Ag^+ or I^- is in or on the solid phase the surface charge density can be calculated from the expression,

$$\sigma = e^{-}(\Gamma_{Ag^+} - \Gamma_{I^-}). \tag{4.28}$$

In order to obtain the surface excesses from the adsorption data the surface area of the silver iodide in the cell must be known. This can be determined with reasonable certainty by at least two independent methods, but these will not be described here. Unless an effectively swamping concentration of inert electrolyte is present (e.g. the 10^{-3} M KNO_3 mentioned earlier) some of the adsorbed Ag^+ or I^- may be in the aqueous diffuse layer rather than on the solid surface, and the surface charge density will not be accurately given by (4.28). If necessary, a correction can be applied for this phenomenon. Although there is little doubt that the charge estimated by the means set out above gives the correct value of the surface charge density, it is probable that the charge is not localised precisely at the phase boundary. It will be recalled that in the mercury system there is little doubt that the charge is at the surface because mercury is a conductor and its dielectric constant is effectively infinite. Silver iodide, however, is a semi-conductor whose dielectric constant is 13.2. Grimley & Mott (1947) have pointed out that in a crystal such as silver iodide there are lattice defects which take the form of interstitial ions (Frenkel defects) and vacant lattice points (Schottky defects). These defects are equivalent to excess positive or negative charges in the lattice and, in general, it is to be expected that there will be a net charge in the crystal. As the defects are usually mobile the charge takes the form of a diffuse ionic layer which extends inwards from the boundary of the solid phase. When such a crystal is immersed in a solution containing silver or iodide ions an equilibrium is presumably established across the phase boundary, and the relative numbers of interstitial ions and vacant lattice points (and therefore the surface charge) are determined by the composition of the solution. Now if the effective thickness of the diffuse layer in the solid (analogous to the Debye–Hückel reciprocal length parameter) is small, then the charge on the silver iodide will be similar in character to that in the mercury system; it will be located effectively at the phase boundary and there will be little or no variation of potential within the crystal. If, on the other hand, the thickness is large, the charge on the silver iodide will be in the form of a diffuse layer such as exists on the aqueous side of the interface and there may be a considerable variation of potential within the crystal. As to which of these two situations is closest to reality is not readily ascertained. From a practical point of view, the manner in which the charge is distributed within the solid phase does not affect the field or the potential difference in that part of the ionic double layer which lies in the aqueous phase, as can readily be deduced from electrostatic theory. However, the variation of potential in the silver iodide does contribute to the Galvani potential difference between the phases and for this reason tends to obscure the properties of the aqueous part of the double layer. This point is discussed in the following paragraph.

Equation (4.27) may be integrated so as to give the cell e.m.f. relative to that at the zero point of charge (where $\boldsymbol{a}_{\mathrm{Ag^+}} = (\boldsymbol{a}_{\mathrm{Ag^+}})_0$), viz.

$$\Delta E = \frac{RT}{F} \ln \frac{\boldsymbol{a}_{\mathrm{Ag^+}}}{(\boldsymbol{a}_{\mathrm{Ag^+}})_0}. \tag{4.29}$$

This change in the Galvani potential difference between the aqueous and silver iodide phases has been equated to the potential difference across that part of the ionic double layer which lies in the aqueous phase. However, the change in Galvani potential difference in this system can strictly only be related to the ionic double layer potentials by a relationship similar to (2.19), i.e.

$$\Delta E = \Delta(\Delta\varphi)\,(\text{molecular dipoles}) + \Delta(\Delta\varphi)\,(\text{ionic double layer}). \tag{4.30}$$

For ΔE to be equal to $\Delta(\Delta\varphi)$ (ionic double layer) it is clearly necessary for $\Delta(\Delta\varphi)$ (molecular dipoles) to be zero. In other words, the dipole contribution to the Galvani potential difference must remain constant and independent of both the ionic double layer potential and the composition of the aqueous phase. The latter may be true, especially in dilute solutions, but whether or not these conditions hold generally is difficult to assess. Even so there is still the difficulty that the diffuse distribution of charge in the silver iodide may give rise to an appreciable potential change in the solid phase. If this does occur, and at present the situation is not entirely clear, it is obviously difficult to infer anything very definite regarding the magnitude of the potential differences in the aqueous part of the double layer. This point is illustrated in fig. 4.11.

4.6. The silver iodide–electrolyte solution interface: some results and their interpretation. This section will be more brief than that for the mercury system because less is known to the same degree of certainty. As a basis for discussion the results of Lyklema & Overbeek (1961) will be considered.

Some data for the surface charge density as a function of the e.m.f. of the cell (relative to that at the zero point of charge) are shown in fig. 4.12. The foreign electrolyte, which is potassium fluoride, was chosen in the expectation that, as in the mercury system, the fluoride ion would show no tendency to adsorb specifically. The differential capacitances calculated from the slopes of the curves in fig. 4.12 are given in fig. 4.13. The dashed curves indicate the differential capacitance of the diffuse layer as calculated from Gouy–Chapman theory. Thus, from the surface charge density the potential drop across the diffuse layer was calculated

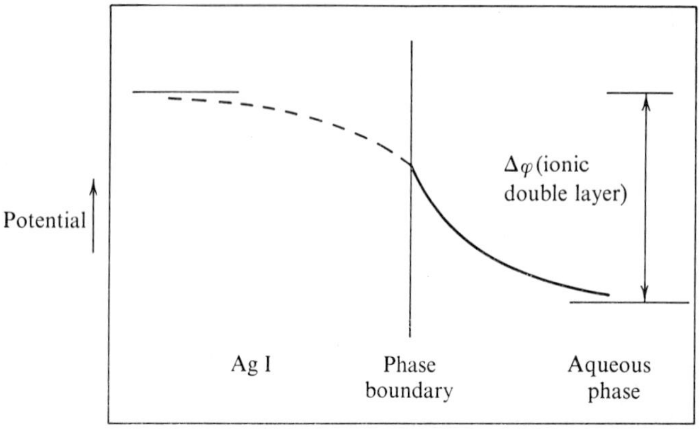

Fig. 4.11. The diffuse ionic double layer at the silver iodide–aqueous solution interface. The dashed curve takes account of the possibility that a diffuse charge may be present in the solid phase.

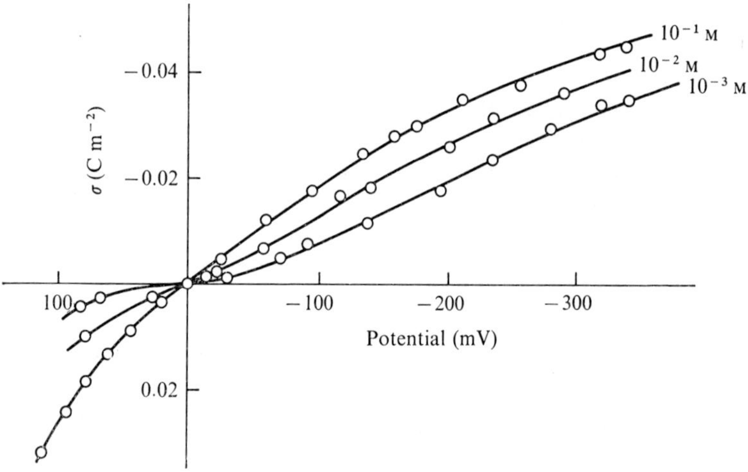

Fig. 4.12. The adsorption of I^- and Ag^+ ions on silver iodide in solutions of potassium fluoride at the concentrations shown. The potential is that of the cell relative to the potential which corresponds to zero charge on the Ag I. (After Lyklema & Overbeek, 1961.)

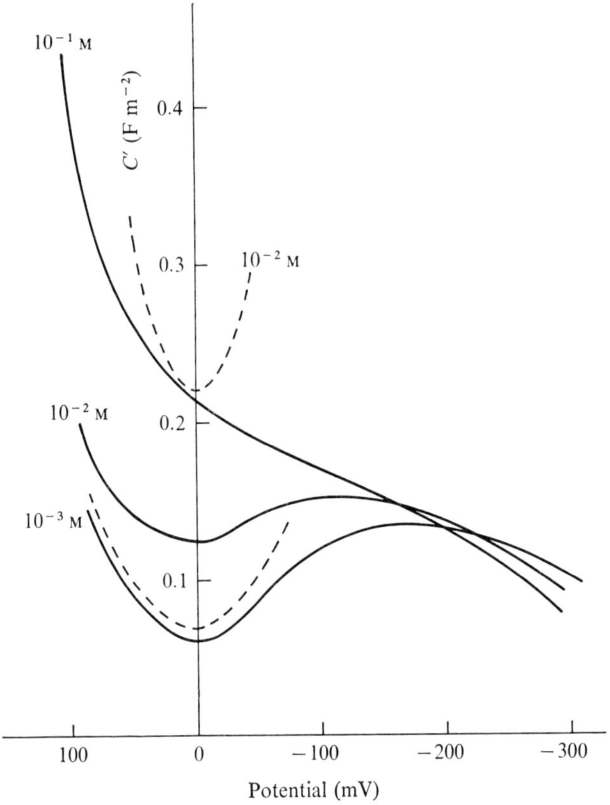

Fig. 4.13. The differential capacitance C' of the double layer on silver iodide in 10^{-3}, 10^{-2} and 10^{-1} M potassium fluoride solutions as a function of the potential. The dashed curves show the differential capacitance according to the Gouy–Chapman theory of the diffuse double layer; the theoretical curve for 10^{-1} M is too high to be included. (After Lyklema & Overbeek, 1961.)

from (2.31) and hence the differential capacitance was obtained by substitution in (4.22). In the lowest electrolyte concentration and near the zero point of charge the agreement between theory and experiment is fairly good, as it was for mercury (fig. 4.5). Although specific adsorption of fluoride ions is discounted, especially when close to the zero point of charge, the contribution of the molecular capacitor formed by the non-specifically adsorbed counter-ions and the interface, to the differential capacitance of the whole double layer will nevertheless usually be significant. This effect is given quantitatively by (4.25), from which it appears that only when the capacitance of the diffuse layer is much less than that of the molecular capacitor, will the latter be negligible. In consequence, in 10^{-3} M KF it is to be expected that, as found experimentally, the capacitance of the whole double layer would be slightly lower than that of the diffuse layer.

At higher potentials and at higher electrolyte concentrations the diffuse layer capacitance becomes larger and the properties of the system become increasingly dominated by the molecular capacitor. It is not unexpected, therefore, that at high negative potentials the differential capacitance becomes largely independent of electrolyte concentration (fig. 4.13). At high positive potentials the capacitance rises sharply and it is supposed, by analogy with the mercury system, that this is attributable to specific adsorption of anions.

For electrolytes other than KF, differences in behaviour are produced by the cation on the negative side of the zero point of charge, and by the anion on the positive side of the zero point of charge. The observed differential capacitances on the negative side decrease in the order

$$NH_4^+ > Rb^+ > K^+ > Na^+ > Li^+$$

and on the positive side in the order

$$NO_3^- > ClO_4^- > F^-.$$

The differential capacitances of the molecular capacitor on silver iodide in the various electrolyte solutions, as calculated from (2.31), (4.22) and (4.25) are shown in fig. 4.14. The resemblance to the comparable result for mercury (fig. 4.6) is notable. The general pattern of results for silver iodide systems is in many respects similar to that for mercury. However, it has been stressed that the interpretation of the silver iodide results is less certain than that for mercury. For example, the possible importance of a diffuse layer in the solid silver iodide was not commented on by Lyklema and Overbeek and neither has it been considered in the above brief description of their work. With this difficulty in particular unresolved, it is probably not profitable in this book to pursue further the reasons for the differences between the mercury and silver iodide systems. It is, however, noteworthy that silver iodide and ice share a similar crystal structure so that some structuring in the water adjacent to the interface may occur. Lyklema has pointed out that this may be a significant factor in the understanding of the properties of the molecular capacitor on silver iodide.

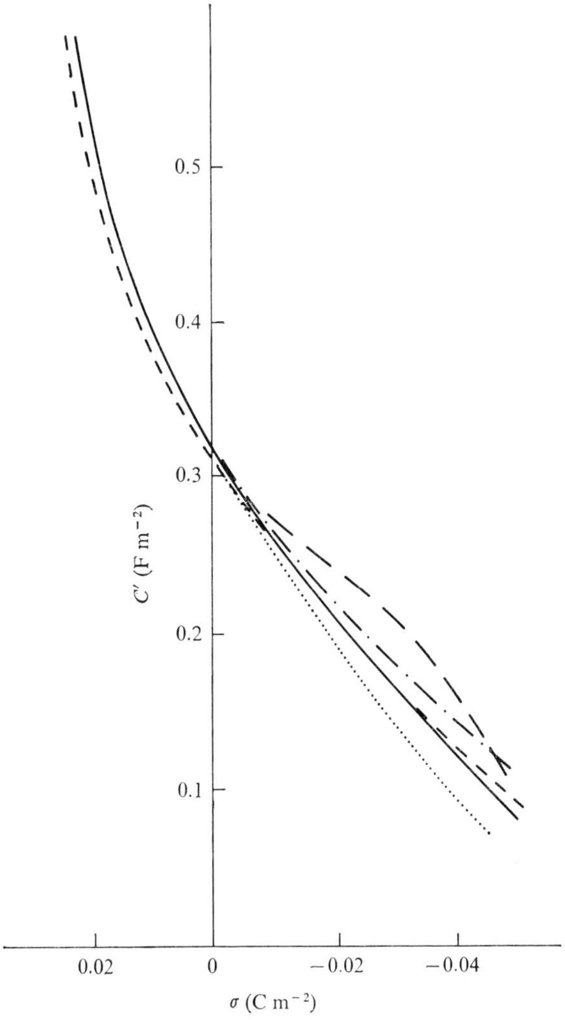

Fig. 4.14. The differential capitance of the molecular capacitor of the double layer on silver iodide in solutions of some monovalent electrolytes.

....	$LiNO_3$
——	KNO_3
----	KF
–·–·–·	$RbNO_3$
— — —	NH_4NO_3

(After Lyklema & Overbeek, 1961.)

References

Grahame, D. C. (1947). *Chem. Rev.* **41**, 441.
Grahame, D. C. (1954). *J. Amer. Chem. Soc.* **76**, 4819.
Grahame, D. C. (1957). *J. Amer. Chem. Soc.* **79**, 2093.
Grahame, D. C. & Whitney, R. B. (1942). *J. Amer. Chem. Soc.* **64**, 1548.
Grimley, T. B. & Mott, N. F. (1947). *Disc. Faraday Soc.* **1**, 3.
Lyklema, J. & Overbeek, J. Th. G. (1961). *J. Colloid Sci.* **16**, 595.
Overbeek, J. Th. G. (1952). *Colloid Science* (ed. H. R. Kruyt), (Elsevier, Amsterdam): vol. 1, p. 160.
Parsons, R. (1954). *Modern Aspects of Electrochemistry* (ed. J. O'M. Bockris & B. E. Conway), (Butterworths, London): p. 103.

5 The solid–gas interface

5.1. Introduction. The surfaces of solids characteristically adsorb gases and vapours. The amount adsorbed by unit mass of the solid depends upon, amongst other things, the pressure of the gas, the temperature of the system and the area which the solid presents to the gas phase. The latter quantity in turn depends upon the state of subdivision of the solid and, more importantly, upon the degree of porosity of the particles. The precise definition of the surface area of a solid is difficult and the experimental determination of its value is subject to considerable uncertainty. It will be recalled that for liquid surfaces it is a relatively simple matter to calculate the number of film molecules in unit area, both for soluble and insoluble films. The treatments of solid–gas adsorption do not normally require an explicit knowledge of the surface area of the solid but rather, the number of molecules needed to complete a monolayer on the sample. For this reason, equations derived from models for solid–gas adsorption are in fact used to calculate the specific surface areas of solids (see §5.12).

Physical adsorption is a dynamic process in that a given molecule does not become irreversibly attached to the surface. Suppose first that adsorbate molecules rebound elastically from a perfectly planar solid surface (a hypothetical situation), then the angles of incidence and reflection will be equal and no exchange of energy between gas and surface will occur. In fact the kinetic energy of the gas molecule will be the same before and after impact. If T_1 is the temperature of impinging gas, T_2' that of rebounding gas and T_2 the temperature of the solid, a coefficient of accommodation, α, can be defined as

$$\alpha = (T_2'-T_1)/(T_2-T_1). \qquad (5.1)$$

In the case discussed above, T_2' is obviously equal to T_1, α is zero and no adsorption takes place. If, however, there is an attraction between gas molecules and solid, the gas molecules remain at the surface for an increased period of time, and it is useful to consider adsorption in terms of this residence time, τ, at the surface. For a simple model,

$$\tau = \tau_0 \exp(q/RT), \qquad (5.2)$$

where τ_0 is the time spent at the surface by a non-adsorbing molecule ($\sim 10^{-13}$ s) and q is the molar energy of adsorption of the gas. If q is small compared to the value of RT, then τ is almost equal to τ_0 and negligible adsorption takes place. As q becomes larger, τ increases and the adsorption is stronger. For example, at 300 K and for $q \sim 20$ kJ mol^{-1} τ is about 3×10^{-10} s. For τ of this magnitude, energy exchange is complete and the value of α is unity.

A more complete account of this particular aspect of adsorption is presented by de Boer (1968).

5.2. Types of system. As attention will be centred on physical adsorption, relevant systems will be those in which specific chemical interaction between absorbent and adsorbate is absent, or almost so.

Carbon, in various forms and from various sources, is a widely used adsorbent. It can be prepared from naturally occurring organic materials (e.g. coconut shell or bone) and by thermal degradation of synthetic polymers. Subsequent heat and chemical treatment (activation) modifies the properties so as to give materials having different surface areas, surface structure and porosity. A form of carbon called Graphon is useful in academic studies since it has a relatively homogeneous surface and is largely non-porous; it usually has a surface area of between 85 and 90 m^{2}g^{-1}. Metal oxides are also extensively used, particularly TiO_2 (anatase, rutile) and Al_2O_3 (alumina); SiO_2 (silica) is also a commonly encountered adsorbent.

Frequently used adsorbates include the rare gases, hydrogen and nitrogen. The latter is probably the most common in practice since it is employed in routine surface area determinations. Other simple gases and vapours which have often been used include n-alkanes, CO_2 and SO_2. For the study of fundamentals of adsorption in terms of molecular interactions, the simplest possible systems are chosen, for example the rare gases and a suitable form of carbon. However, many adsorption studies are made in order to determine the surface areas and porosity of samples (such as those used in industrial processes) and then a much wider range of systems is examined.

5.3. Solid–gas interactions in physical adsorption. The object of this brief discussion is to indicate how the treatment of the forces between two isolated atoms is extended to cover the system of a single adsorbate atom and a solid surface. Effects other than those which originate from the London–van der Waals forces and from the repulsion between electron clouds are not considered. A much fuller account is given, for example, by Young & Crowell (1962*a*).

The net interaction energy, $U(r)$, between two isolated atoms may be assumed to have the form

$$U(r) = -Cr^{-6}+Br^{-n}, \tag{5.3}$$

where r is the separation of the atoms and C, B and n are constants. $-Cr^{-6}$ is the dispersion energy (i.e. energy of attraction) between the atoms, and Br^{-n} is the energy of repulsion due to the mutual proximity of the electron clouds. If n is chosen as 12, $U(r)$ becomes the familiar Lennard-Jones 6–12 potential, although values of n between 9 and 14 often lead to a reasonable fit of data.

The interaction between an adsorbate atom (often called an *adatom* in this context) and a solid surface can be taken as the sum of the interaction between the adatom and the solid atoms. Thus, the dispersion energy Φ_D between adatom and solid is given by

$$\Phi_D = -\sum_i Cr_i^{-6}, \tag{5.4}$$

where i refers to the ith solid atom and r_i is the distance of the adatom from i. London (1930) derived an expression for Φ_D by replacing the summation (5.4) by a volume integral to give

$$\Phi_D = -\int Cr^{-6}\rho_N \,\mathrm{d}v, \tag{5.5}$$

where ρ_N is the number of solid atoms per unit volume and $\mathrm{d}v$ is the volume element of the solid at a distance r from the adatom. For a solid of constant density (5.5) leads to the expression

$$\Phi_D = -\frac{\rho_N \pi C}{6z^3}, \tag{5.6}$$

where z is the perpendicular distance of the adatom from the surface.

It should be pointed out that the use of (5.5) rather than (5.4) is valid only for values of z much larger than the interatomic spacing in the solid. Since the equilibrium separation of adatom and solid surface is of the same order as the interatomic spacing in the solid, (5.5) does not lead to a quantitative picture of the attraction between adatom and adsorbent at this separation. Nevertheless a useful qualitative description is obtained.

A similar procedure to that outlined above can be followed to obtain an expression for the energy of repulsion Φ_R between the adatom and solid. It can be shown that, for $n = 12$ in (5.3)

$$\Phi_R = \frac{2\pi\rho_N B}{90z^9}. \tag{5.7}$$

Thus, the net interaction energy $\Phi(z)$ obtained by adding (5.6) and (5.7) is

$$\Phi(z) = -\frac{\pi\rho_N C}{6z^3} + \frac{2\pi\rho_N B}{90z^9}. \tag{5.8}$$

In terms of the equilibrium separation z_0 of solid and adatom, (5.8) may be written

$$\Phi(z) = -\frac{\pi\rho_N C}{6z_0^3}\left[\left(\frac{z_0}{z}\right)^3 - \frac{1}{3}\left(\frac{z_0}{z}\right)^9\right] \tag{5.9}$$

since when $z = z_0$, $\mathrm{d}\Phi(z)/\mathrm{d}z = 0$.

Thus, an interesting conclusion that can be drawn from the simple outline above, is that the assumption of a 6–12 potential function between isolated atoms leads to a 3–9 potential function between an adatom and a plane solid surface. In other words, the attractive force falls off much less rapidly with distance than for two isolated atoms. In addition, it is a simple matter to show that the equilibrium separation of adatom and solid is smaller than for the two isolated atoms.

The expression (5.8) will be used later (in §5.14) for the derivation of an isotherm for multilayer adsorption on non-porous solids.

5.4. Surface tension and surface free energy of solids. It will be recalled (§1.2) that for one component liquids, if the Gibbs surface is placed such that $\Gamma^\sigma = 0$, the surface tension is equal to the (specific excess) surface free energy. For solids, however, the two quantities are not in general equal. When a fresh solid surface is formed the atoms at the surface may take a considerable time to come to their equilibrium positions, and therefore the new surface is in a non-equilibrium state.

It is possible, in principle, to extend the area of a surface in two ways (Benson & Yun, 1967). One way is to create fresh surface having the same properties as the original, and the other is to stretch the surface already present. This can easily be understood by referring to simple processes involving liquid systems. Under most experimental circumstances the extension of the surface of a pure liquid or of a liquid with a soluble monolayer present will result in a surface having the same surface tension as the original. The extension of an isolated segment of a thin soap film, would, however, be an example of the second type of surface formation. Here there is effectively no reservoir of material to maintain the surface at constant composition, and so the stress in the surface changes with the extension. The same holds for the compression or expansion of an insoluble monolayer, where the surface pressure varies with the film area.

With the above discussion in mind, consider the work, $\mathrm{d}W$, done in the extension of the surface of a one-component isotropic solid by an amount $\mathrm{d}\mathcal{A}$. Suppose that the work is done against a force per unit length in the surface (i.e. surface tension) γ^s. Then $\mathrm{d}W$ is given by $\gamma^s \mathrm{d}\mathcal{A}$, and also by $\mathrm{d}(\mathcal{A}A_\sigma)$ where A_σ is the specific excess surface free energy. Thus,

$$\gamma^s \mathrm{d}\mathcal{A} = \mathrm{d}(\mathcal{A}A_\sigma)$$

so that

$$\gamma^s = \frac{\mathrm{d}}{\mathrm{d}\mathcal{A}}(\mathcal{A}A_\sigma)$$

and hence

$$\gamma^s = A_\sigma + \mathcal{A}\left(\frac{\mathrm{d}A_\sigma}{\mathrm{d}\mathcal{A}}\right). \tag{5.10}$$

It is clear that the surface tension γ^s is only equal to A_σ when

$$\mathrm{d}A_\sigma/\mathrm{d}\mathcal{A} = 0.$$

This is true, as explained above, for a one component liquid. For a freshly formed solid surface, however, which is not at equilibrium, this is not true and indeed the magnitudes of A_σ and $\mathcal{A}(\mathrm{d}A_\sigma/\mathrm{d}\mathcal{A})$ may be comparable.

The surface tension γ^s has been defined directly in terms of the force acting along unit length in the surface. It is worth noting, however, that some authors (e.g. Johnson, 1959) refer to the quantity

$$\gamma = (\partial A/\partial \mathcal{A})_{T,V,n}$$

as the surface tension of a solid. Nevertheless, for solids, γ cannot be equated numerically to a tension, except for an equilibrium surface, and Gibbs did not give a name to this quantity. It is worth recalling here that even for equilibrium surfaces, surface tension and surface free energy are not equal if adsorption is occurring in the system (see (1.26)).

Readers interested in the procedures involved in the experimental and theoretical estimation of surface free energies and surface tensions are referred to Adamson (1967) and Gregg (1965*a*) or, for a more advanced account, to Benson & Yun (1967). It is sufficient to note here that experimental methods are often complex and the results can often be subject to considerable uncertainty. This situation can be contrasted with that for liquids where surface tensions can usually be determined with a high degree of precision (§3.3).

5.5. Surface pressure in solid–gas systems. The surface pressure, π, exerted by a gas adsorbed on a solid surface is defined as

$$\pi = \gamma_0 - \gamma, \tag{5.11}$$

where the γs are defined as in (1.16) and the subscript 0 refers to the solid–vacuum interface. Thus, π is defined in an entirely analogous way to that for liquid–vapour and liquid–liquid interfaces (§1.3).

In a single gas–solid system the Gibbs adsorption equation (1.40) may be written (see §1.10)

$$-\mathrm{d}\gamma = RT\Gamma^s \mathrm{d} \ln p, \tag{5.12}$$

where Γ^s is the surface concentration and p the pressure of the adsorbate. Integration of (5.12) yields

$$\gamma_0 - \gamma = \pi = RT \int_{p=0}^{p} \Gamma^s \mathrm{d} \ln p. \qquad \blacktriangleleft\mathbf{(5.13)}$$

The direct determination of γ_0 and γ is not usually possible, but it can be seen from (5.13) that $\gamma_0 - \gamma$ i.e. the surface pressure, can be obtained indirectly. If m grams of gas of molecular weight M are adsorbed on 1 gram of solid of surface area Σ ($\mathrm{m^2\,g^{-1}}$), then

$$\Gamma^s = m/M\Sigma \text{ mol m}^{-2}$$

and (5.13) may be written

$$\pi = \frac{RT}{M\Sigma} \int_{p=0}^{p} m \,\mathrm{d} \ln p. \tag{5.14}$$

The surface pressure corresponding to any equilibrium gas pressure between 0 and p may be obtained according to (5.14) by graphical integration of a plot of m versus $\ln p$.

The area, a, occupied by a mole of gas on the surface is

$$a = \frac{\Sigma M}{m} \text{m}^2. \tag{5.15}$$

Thus, (5.14) and (5.15) may be used to convert experimental results for the adsorption of gas on a solid, into π and a values. Therefore, although γ is not usually measurable for solids, the applicability of surface equations of state of the form of (1.49) may still be investigated.

5.6. Experimental determination of adsorption. A wide variety of apparatus and techniques has been used to determine the extent of adsorption of gases on to solid surfaces. Ross & Olivier (1964) have reviewed some of the available procedures. Broadly speaking, the measurements fall into one of two categories; either the *volume* of gas adsorbed is determined manometrically, or a gravimetric method is used where the *mass* adsorbed is determined directly. The volumetric method is probably the most widely used. Various refinements are required for systems in which the gas pressure is particularly high or low. An outline will be given here of a simple apparatus which can be used for the measurement of the adsorption of, say, nitrogen on to most adsorbents at about -196 °C (the boiling point of liquid nitrogen at atmospheric pressure, liquid nitrogen being used as a means of thermostatting the adsorbent sample). The use of low temperatures, as opposed to say room temperature, greatly enhances the fairly weak adsorption and under these conditions the measurements become much easier. As will be seen later, nitrogen adsorption at -196 °C is commonly measured in the routine determination of the surface areas of adsorbents. Fig. 5.1 shows an apparatus described by Emmett (1942) which is suitable for such measurements.

The calibration of the apparatus involves the determination of the volumes of the gas burette bulbs and the volume enclosed by taps T_1 and T_3 and the mark M on the manometer. When pressure readings are taken on the manometer it is always necessary to adjust the mercury in the left-hand arm to the mark M by appropriate use of the mercury reservoir attached to the manometer.

The adsorption isotherm is obtained in the following way. The adsorbent is placed in the sample bulb, which is detachable. Then, with taps T_1 and T_2 open to the main vacuum line, and T_3 open to the sample, the apparatus and sample are outgassed until a suitably low pressure is registered on the pressure gauge. Taps T_1, T_2 and T_3 are then closed and an amount of helium is admitted via T_1 and the pressure is measured by means of the manometer. T_3 is then opened in order to allow helium into the sample bulb. The pressure is again measured. Since helium is not significantly adsorbed it is now possible to calculate the 'dead space' up to T_3 (i.e. the volume not occupied by the adsorbent). The helium is now withdrawn from the apparatus and a dose of nitrogen admitted, T_3 being closed. Knowing the pressure and volume of nitrogen at the temperature of the water jacket, the number of moles n_1 present can be

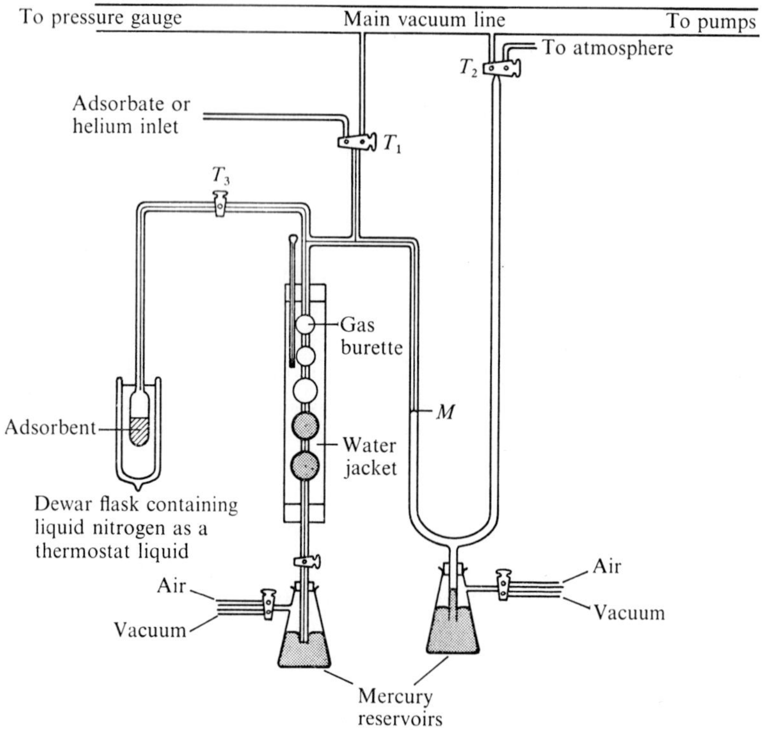

Fig. 5.1. Gas adsorption apparatus (after Emmett, 1942).

calculated. Tap T_3 is now opened so that the gas enters the sample bulb and is adsorbed on to the sample. When the system has attained equilibrium the pressure p_1 is measured and the number of moles n_2 of nitrogen in the gas phase can be calculated. $(n_1 - n_2)$ moles of nitrogen have therefore been adsorbed at pressure p_1. The pressure is now increased by forcing mercury into the gas burette from the mercury reservoir and further readings are taken. The adsorption corresponding to these pressures is calculated. The desorption isotherm is determined by the reversal of the process.

Adsorption may be determined gravimetrically by the use of a microbalance which operates within the vacuum system. A simple balance consists of a calibrated quartz spiral from which hangs a pan containing the adsorbent. The amount of gas adsorbed is determined from the extension of the spiral. The balance (fig. 5.2) is connected to a suitable dosing device and manometer.

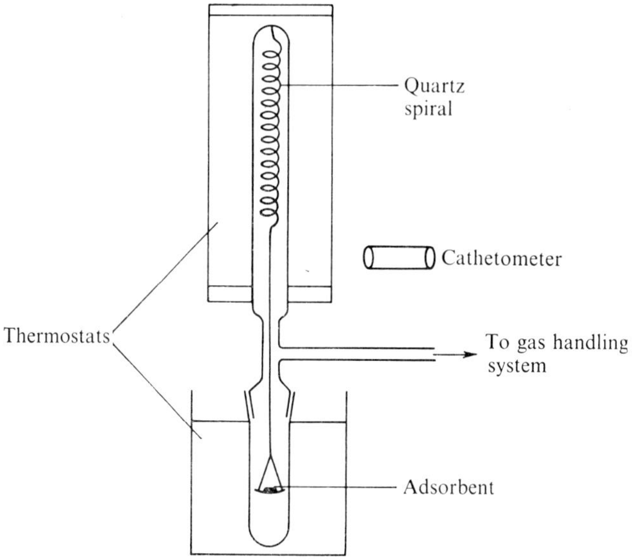

Fig. 5.2. Adsorption balance.

Adsorption isotherms are normally presented as plots of the amount adsorbed versus the equilibrium pressure p, or relative pressure p/p^0, where p^0 is the saturated vapour pressure. The amount adsorbed is usually expressed as the volume v of gas referred to STP, the mass m, or the mass per gram of adsorbent m/m_A, where m_A is the mass of the adsorbent sample.

5.7. Classification of experimental isotherms. Experimental adsorption isotherms recorded in the literature have a wide variety of forms. Brunauer, Deming, Deming & Teller (1940) made a very useful empirical classification of data obtained for temperatures below the critical temperature of the adsorbate. Most of the observed isotherms fall into one of the five categories depicted in fig. 5.3. The types are referred to by the numbers shown in the figure.

Type I isotherms rise sharply at low relative pressures and reach a plateau. The shape is consistent with the formation of a monolayer upon which no further adsorption takes place. It is now known, however, that type I isotherms very often result when the adsorbent is microporous (§5.16). In this case the limiting value of the adsorption reflects

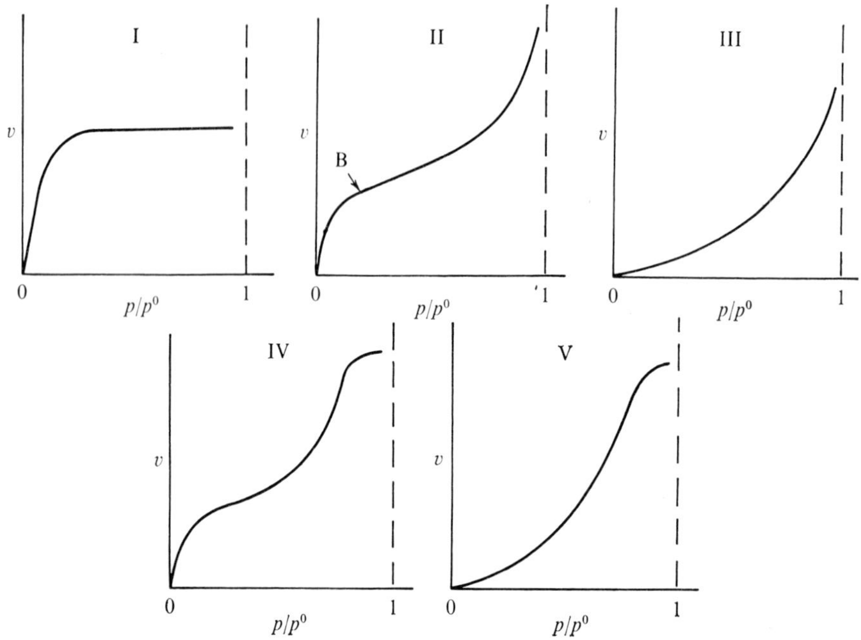

Fig. 5.3. Classification of isotherms according to Brunauer *et al.* (1940). (The point *B* in the isotherm of type II is at the beginning of the linear portion of the isotherm; it is referred to in §5.12.)

the filling of the micropores rather than the completion of a monolayer. Many early experiments were performed using charcoal and chabazite (a complex hydrated sodium and calcium silicate), which are porous adsorbents. These solids frequently give type I isotherms with simple adsorbates, and it was erroneously concluded that monolayer adsorption was very common. In fact, multilayer adsorption is much more frequently encountered in gas–solid systems.

Types II to V isotherms all result from multilayer adsorption. Type II is similar to type IV except that the latter tends to level off at high relative pressures. A similar relationship holds between types III and V.

Isotherms of types II and III are found for multilayer adsorption on non-porous solids. Type II isotherms are very commonly encountered but those of type III are rare, bromine vapour on silica gel and water vapour on graphite being two examples. Isotherms of types IV and V are characteristic of adsorption on porous solids (§5.18).

5.8. Agreement between experimental and theoretical isotherms. It is appropriate, before discussing specific models for adsorption, to point out some of the difficulties in accounting for experimental results. Many surface equations of state and adsorption isotherms have been advanced for gas–solid systems. As Steele (1967), for example, has commented, different models often lead to theoretical isotherms which have similar mathematical form and thus can be made to fit experimental data equally well. For this reason, it is quite unsatisfactory to compare experiment and theory using data obtained at a single temperature. It is desirable to compare the thermodynamic functions predicted from a certain model with those obtained from the determination of adsorption isotherms at various temperatures (see §5.22).

There are nevertheless relatively simple equations based on somewhat inadequate models, which are of considerable use to practising surface chemists. In the description which follows, no attempt is made to give the details of recent work, which is both extensive and complex. Rather, an outline is given of some of the simpler ideas which underlie the study of gas–solid adsorption. Those wishing to pursue the subject further are recommended to consult Young & Crowell (1962), Ross & Olivier (1964), and Flood (1967).

5.9. Monolayer adsorption. The study of monolayer adsorption still presents many problems and the present state of the subject has been reviewed by Steele (1967). Before experimental results can be discussed in terms of a theoretical isotherm it is necessary to have some idea as to the nature of the adsorbed film. Thus, it is helpful to know whether the film is monomolecular or multimolecular and if it is likely to be localised or non-localised, although in physically adsorbed films it is unlikely that true localisation exists. It is also probably true that completely non-localised films are rare (see §1.5). It is a complicated matter to deal theoretically with films which exhibit intermediate behaviour, however, and only the extreme cases of complete localisation and complete non-localisation are discussed here. In addition, consideration of the effects of surface heterogeneity, which is often encountered in practice, is omitted.

Both localised and non-localised adsorption isotherms reduce to the so-called Henry's law (1.60) for sufficiently dilute films. For the present purposes, (1.60) may be written in the form

$$p = K'\theta, \tag{5.16}$$

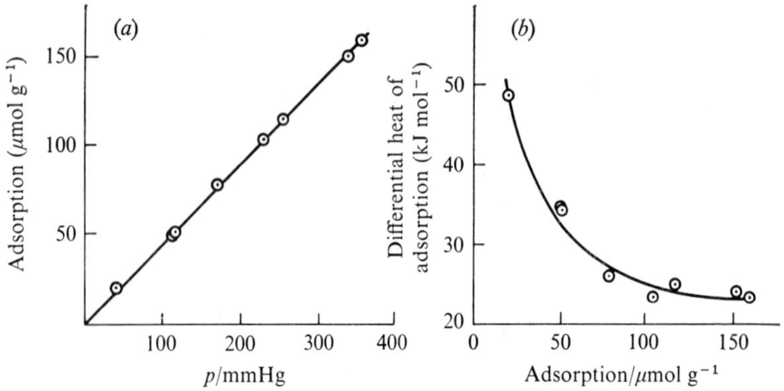

Fig. 5.4. (*a*) Adsorption isotherm for CO_2 on silica gel at 75 °C. (*b*) Calorimetric differential heats of adsorption of CO_2 on silica gel at 75 °C. (Results taken from Kälberer & Schuster, 1929.)

where K' is a constant and θ the surface coverage. Equation (5.16) predicts a linear relation between θ and adsorbate pressure. This linearity corresponds to the initial rise in isotherms of type I. The equation is only likely to be obeyed for very small values of θ (§1.10). Examples of linear, low-coverage isotherms are to be found in the literature, but it cannot be supposed with confidence that the model assumed in the derivation of (5.16) is realistic for the experimental systems. Fig. 5.4 shows results obtained by Kälberer & Schuster (1929) for the adsorption of CO_2 on silica gel at 75° C. The linearity of the isotherm (fig. 5.4(*a*)) indicates the constancy of K' in (5.16). However, the calorimetric differential heats of adsorption (fig. 5.4(*b*)) are by no means constant over the same range of coverage (see §5.22). It is possible that this variation was due to surface heterogeneity of the adsorbent.

The Langmuir equation

$$p = K_3 \frac{\theta}{1-\theta} \tag{1.87}$$

has been widely used to describe the adsorption of gases on solids. $\theta = \dot{a}_0/\dot{a}$ where $\dot{a}_0$ is the co-area of the adsorbate molecule (§1.12). Strictly, for a localised model, the area of an adsorption site should be used but this is not normally known. θ may also be written in terms of the volume, v, of gas adsorbed so that $\theta = v/v_m$. The quantity v_m is the volume of gas required to cover the surface with a complete monolayer only. If the adsorption is referred to 1 gram of solid, v_m is termed the *monolayer capacity*. Equation (1.87) is often written in the form

$$v = \frac{v_m bp}{1+bp}, \tag{5.17}$$

where the constant b is equal to $1/K_3$. The nature of b is evident from (1.88).

It is clear from (5.17) that at low pressures

$$v \simeq v_m bp \tag{5.18}$$

which is equivalent to (5.16). At high pressures, where $bp \gg 1$ (5.17) becomes

$$v \simeq v_m. \tag{5.19}$$

This equation indicates that at high pressures the adsorption tends to a limiting value (v_m), for which the monolayer is complete. Thus the Langmuir equation can represent isotherms of type I.

The applicability of the equation to experimental data can be tested by converting (5.17) to a suitable linear form such as

$$\frac{p}{v} = \frac{p}{v_m} + \frac{1}{v_m b}, \tag{5.20}$$

i.e. a plot of p/v versus p should be a straight line. Clearly, the constants v_m and b may be obtained from such a plot. It has already been pointed out (§5.7) that adsorption data for many systems follow the form of the Langmuir equation although the adsorbed films do not approximate to the model. Fig. 5.5 shows such an example. Fig. 5.5(*a*) is the adsorption isotherm for nitrogen on a microporous carbon at −196 °C and fig. 5.5(*b*) is the linear plot according to (5.20) (Wayman, 1967). Similarly, conformity of experimental results with (1.76) for non-localised films would not necessarily confirm the validity of the model for a particular system.

Equations have been derived to account for the presence of interactions within a film (§§1.11 and 1.13). An example is (1.98) for a localised monolayer. The so-called Bragg–Williams approximation is used in the derivation, in which it is assumed that the distribution of adsorbed molecules over the adsorption sites is random, even though lateral attractions are present. Such a treatment is analogous to the zeroth approximation for regular solutions (Guggenheim, 1952). An

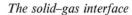

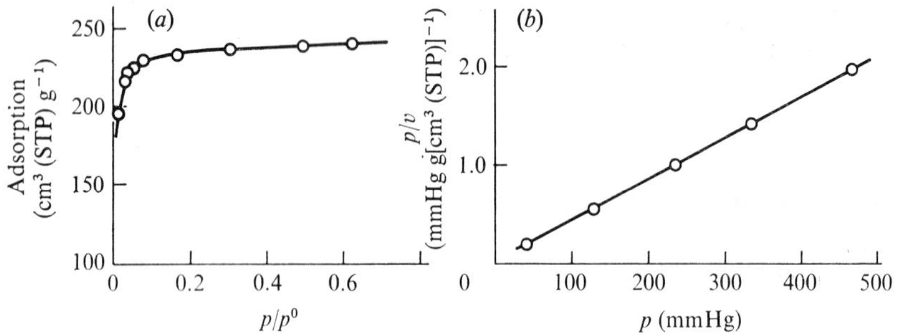

Fig. 5.5. (*a*) Adsorption isotherm for N_2 on a microporous carbon at −196 °C. (*b*) Same results as in (*a*) plotted according to (5.20). (Results from Wayman, 1967.)

improvement on the assumption of complete randomness is embodied in the quasi-chemical approach, which leads to the equation

$$p = K_4 \frac{\theta}{1-\theta}\left(\frac{2-2\theta}{\beta+1-2\theta}\right)^c, \tag{5.21}$$

where $\beta = \{1-4\theta(1-\theta)(1-\exp[-V/kT])\}^{\frac{1}{2}}$.

The symbols have the same significance as in (1.98). Since this isotherm is rather complex, it has only a limited practical value.

The isotherm (1.79) has been derived for non-localised monolayers with lateral interactions. However, as both (1.79) and (1.98) contain several disposable constants these equations could at least to a first approximation, be made to fit the same experimental data. It is therefore impossible by the use of these equations alone to ascertain whether or not a film is localised. It is necessary to compare the constants with values obtained independently.

5.10. Condensation in monolayers. It might intuitively be expected that if a monolayer becomes sufficiently concentrated, two-dimensional condensation will take place. Indeed, both (1.79) and (1.98) predict this for temperatures below the two-dimensional critical temperature. Such condensation has been experimentally observed. In fig. 5.6 are shown adsorption isotherms for the system ethane on {100} surfaces of sodium chloride at several temperatures between 122.9 and 147.6 K (Ross & Clark, 1954). The vertical sections correspond to the condensation. It can be seen that the two-dimensional critical temperature T_{2c} lies between 131.2 and 136.0 K. To show the similarity with condensation

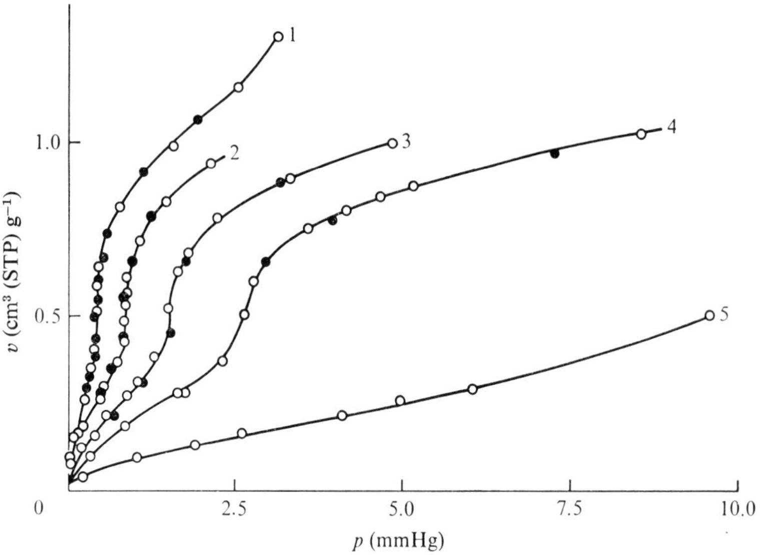

Fig. 5.6. Low pressure isotherms for ethane on {100} surfaces of sodium chloride. Open circles represent adsorption, and closed circles desorption points. Curve 1, 122.9 K; 2, 127.6 K; 3, 131.2 K; 4, 136.0 K; 5, 147.6 K. (After Ross & Clark, 1954.)

in insoluble monolayers at the liquid–vapour interface (cf. fig. 3.17) the data of fig. 5.6 can be converted to π against $\dot{a}$ curves. The surface pressure is calculated in the manner described in §5.5, and the results are shown in fig. 5.7.

Hill (1946*a*) has derived relationships between the two-dimensional critical constants (area, pressure and temperature) for a non-localised film, and the three-dimensional critical constants. It was found that T_{2c} should theoretically be half the three-dimensional value of T_{3c}, which is born out reasonably well in practice. For example, in the case of ethane adsorbed on NaCl it is found that $T_{2c}/T_{3c} = 0.43$ in moderate agreement with the predicted value of 0.5.

5.11. Multilayer adsorption: introduction. As physical adsorption is a consequence of van der Waals forces, and these forces are also operative in the liquefaction of gases, adsorption need not stop on the completion of a monolayer. In fact, the formation of multilayers, which are essentially liquid by nature, is extremely common. The theory of the liquid state is complex and is at present incomplete. Thus, the problem of

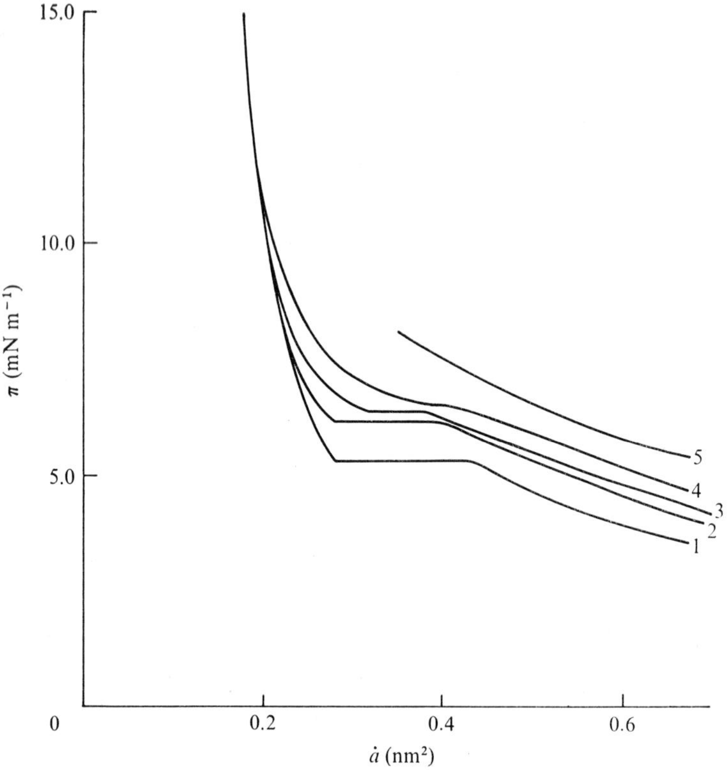

Fig. 5.7. Pressure-area curves for ethane on sodium chloride. Curve numbers as in Fig. 5.6. (After Ross & Clark, 1954.)

treating rigorously a liquid film in the potential field of a solid adsorbent has not been solved, and simplifications must be introduced. If the adsorbent is porous, which in practice is common, a modified approach is necessary and it will be convenient to treat such cases separately in later sections. Only the adsorption of single gases is considered.

5.12. The BET theory. This theory was introduced by Brunauer, Emmett & Teller (1938). Hill (1952) has summed up the status of the theory. He points out that the theory was the first of its kind and that it has stimulated much subsequent research in solid–gas adsorption. It is still very widely used, especially in the estimation of the specific surface areas of solids and, while it is not quantitatively correct, it gives a useful qualitative description of the adsorption process.

The original derivation of the BET equation was based on kinetic arguments and was an extension of Langmuir's treatment of monolayer adsorption. However, as for the Langmuir equation (§ 1.12), a statistical derivation is preferable since the nature of the model is seen more clearly and it is not necessary to assume any kinetic mechanism. In addition, the constants of the equation are given in terms of molecular parameters. The derivation below closely follows that given by Hill (1946*b*, 1952).

It is assumed that the adsorption is localised. Unit area of the adsorbent surface contains N_s equivalent adsorption sites of which N_1 are occupied by adsorbate molecules. The N_1 molecules thus constitute the first layer of adsorbate. If the total number of adsorbate molecules at the surface is N then $(N-N_1)$ molecules are in subsequent layers. Each first layer molecule acts as a potential adsorption site for a second layer molecule, which in turn acts as a site for a third layer molecule and so on, there being no restriction on the total number of molecules in any given stack. It is supposed that molecules in the second and higher layers have the same partition function and energy as in the bulk liquid. The molecules in the first layer will, in general, have a different partition function. It is further assumed that the stacks do not interact energetically.

First layer molecules are characterised by the potential energy $-V_1$ (relative to zero at infinite separation from the surface) and molecules in all subsequent layers by $-V_l$. The partition function for a molecule in the first layer is then (see § 1.12)

$$\mathfrak{z}_1(T) = j_1^{\text{ads}}(T) \exp\left(\frac{V_1}{kT}\right). \tag{5.22}$$

For a molecule in any other layer, the partition function $\mathfrak{z}_l(T)$ is

$$\mathfrak{z}_l(T) = j_l^{\text{ads}}(T) \exp\left(\frac{V_l}{kT}\right). \tag{5.23}$$

The complete partition function Z_1^{ads} for the first layer is (cf. (1.81))

$$Z_1^{\text{ads}} = \frac{N_s!}{N_1!(N_s-N_1)!}[\mathfrak{z}_1(T)]^{N_1}. \tag{5.24}$$

It will be recalled (§ 1.12) that $N_s!/N_1!(N_s-N_1)!$ is the number of distinguishable ways of arranging N_1 molecules over N_s sites. The number of distinguishable ways in which the $(N-N_1)$ remaining molecules may be distributed on top of the N_1 first layer molecules, there being no restriction on the number of molecules in any stack, is $(N-1)!/(N-N_1)!(N_1-1)!$. Thus, the partition function Z_l^{ads}, for the $(N-N_1)$ molecules is given by

$$Z_l^{\text{ads}} = \frac{(N-1)!}{(N-N_1)!(N_1-1)!}[\mathfrak{z}_l(T)]^{N-N_1}. \tag{5.25}$$

The complete partition function, Z^{ads}, for the system is

$$Z^{\text{ads}} = \sum_{N_1=1}^{n} Z_1^{\text{ads}} Z_l^{\text{ads}}, \tag{5.26}$$

where $n = N$ (the total number of molecules) if N is less than N_s and $n = N_s$ if the total number of molecules exceeds the number of sites. If unity is neglected as compared to N and N_1 then (5.26) may be written

$$Z^{\text{ads}} = \sum_{N_1=1}^{n} \left\{ \frac{N!N_s!}{(N-N_1)!(N_s-N_1)!(N_1!)^2}[\mathfrak{z}_1(T)]^{N_1}[\mathfrak{z}_l(T)]^{N-N_1} \right\}. \tag{5.27}$$

It is a sufficiently good approximation for practical purposes to set $\ln Z^{\text{ads}}$ equal to the logarithm of the largest term in the sum. The value of N_1 which corresponds to the maximum term is obtained from

$$\frac{\partial \ln Z_1^{\text{ads}} Z_l^{\text{ads}}}{\partial N_1} = 0. \tag{5.28}$$

Hence it can be shown that

$$(N-N_1)(N_s-N_1) = N_1^2 \frac{\mathfrak{z}_l(T)}{\mathfrak{z}_1(T)} = \frac{j_l^{\text{ads}}(T)}{j_1^{\text{ads}}(T)} N_1^2 \exp\left[\frac{V_l-V_1}{kT}\right]. \tag{5.29}$$

The chemical potential, μ, of the adsorbed molecules is obtained by noting that their Helmholtz free energy A^{ads} is given by

$$A^{\text{ads}} = -kT \ln Z^{\text{ads}} = -kT \ln Z_1^{\text{ads}} Z_l^{\text{ads}}. \tag{5.30}$$

Then

$$\mu = \left(\frac{\partial A^{\text{ads}}}{\partial N}\right)_T = kT\left(\frac{\partial \ln Z_1^{\text{ads}} Z_l^{\text{ads}}}{\partial N}\right)_T. \tag{5.31}$$

In both (5.30) and (5.31), $\ln Z_1^{\text{ads}} Z_l^{\text{ads}}$ is the logarithm of the largest term in the summation in (5.27). From (5.31) it can be shown that

$$\frac{\mu}{kT} = \ln\left(\frac{N-N_1}{N}\right)\frac{1}{\mathfrak{z}_l(T)}. \tag{5.32}$$

The chemical potential of the molecules in the ideal bulk gas, which is in equilibrium with the adsorbed phase, is given by (see (1.59))

$$\frac{\mu}{kT} = \ln p + \ln\left\{\frac{h^3}{kT(2\pi mkT)^{\frac{3}{2}} j^g(T)}\right\} = \ln p + \alpha, \tag{5.33}$$

where α is constant at constant T. Clearly $kT\alpha$ is a standard chemical potential for a molecule in the gas phase.

According to the model, molecules in all but the first layer have the same partition function j_l^{ads} and energy $-V_l$ as molecules in the pure liquid. Thus, the chemical potential $\mu^{0,l}$ of molecules in pure liquid adsorbate is given by

$$\frac{\mu^{0,l}}{kT} = -\ln \mathfrak{z}_l(T). \tag{5.34}$$

This can be understood by reference to (5.25). When N_1 is set equal to zero all the (N) molecules are in the higher layers and, according to the model, equivalent to molecules in pure liquid. Then for $N \gg 1$, Z_l^{ads} becomes $[\mathfrak{z}_l(T)]^N$, and (5.34) results.

When a liquid is in equilibrium with its saturated vapour at pressure p^0 it follows from (5.33) and (5.34) that

$$-\ln \mathfrak{z}_l(T) = \ln p^0 + \alpha. \tag{5.35}$$

Therefore, from (5.32), (5.33) and (5.35),

$$\frac{N - N_1}{N} = \frac{p}{p^0} = x, \tag{5.36}$$

where x is the relative pressure. Finally, the combination of (5.36) and (5.29) yields the well known BET equation

$$\frac{N}{N_s} = \frac{cx}{(1-x)(1-x+cx)}, \tag{5.37}$$

◄**(5.37)**

the constant c being given by

$$c = \frac{j_1^{\text{ads}}(T)}{j_l^{\text{ads}}(T)} \exp\left[\frac{V_1 - V_l}{kT}\right].$$

The term N/N_s is usually expressed as v/v_m, where v and v_m have the same significance as in §5.9.

Some of the shortcomings of the BET model are apparent. The assumption of localised adsorption in all the layers is not obviously in accord with the supposition that the film (excluding the first layer) is liquid. Also, the assumption that the stacks of molecules do not interact energetically is unrealistic. It means that molecules in the 'liquid' layer have only two interacting nearest neighbours, whereas in reality a molecule in a bulk liquid would have up to twelve nearest neighbours. Other criticisms of the BET theory are given, for example, by Young & Crowell (1962*b*).

In spite of these inadequacies the BET theory is very useful in a qualitative sense and isotherms of type II and type III (fig. 5.3) are described. When the constant $c \gg 1$, a type II isotherm is predicted. Here, adsorption in the first layer is strong relative to the adsorption in higher layers, as will be appreciated from the expression for c, so that the first layer is almost completed before higher layers are formed. This accounts for the formation of the 'knee' in the isotherm at low values of p/p^0. As expected, for low pressures (5.37) reduces approximately to the Langmuir equation, i.e.

$$\frac{v}{v_m} \simeq \frac{cx}{1+cx}. \tag{5.38}$$

Equation (5.38) may be compared with (5.17) when it is seen that $c = bp^0$. For small c (say 0.1) (5.37) yields an isotherm of type III. The change from type II to type III occurs for a value of $c = 2$, when the point of inflexion in the isotherm coincides with the origin.

As indicated in §5.7, isotherms of type II are very commonly encountered in practice and it is usually found that the BET equation will fit experimental data for values of p/p^0 from about 0.05 to 0.35. Equation (5.37) may be cast in the form

$$\frac{p}{v(p^0-p)} = \frac{1}{v_m c} + \frac{(c-1)p/p^0}{v_m c} \qquad \blacktriangleleft \textbf{(5.39)}$$

so that if the BET equation represents the data, the plot of $p/v(p^0-p)$ against p/p^0 is linear, and v_m and c may be obtained from the slope and the intercept of the line.

The surface area of an adsorbent may be estimated from the value of v_m. The most commonly used adsorbate in the adsorption method for area determination is nitrogen. This is normally adsorbed on the solid sample at -196 °C (the boiling point of N_2). The value of c for nitrogen on many adsorbents is high and a well defined 'knee' is formed in the resulting type II isotherm. In fact it is often possible to obtain an approximate value for v_m merely by inspection of the isotherm, since it has been found that the point B (fig. 5.3) frequently corresponds

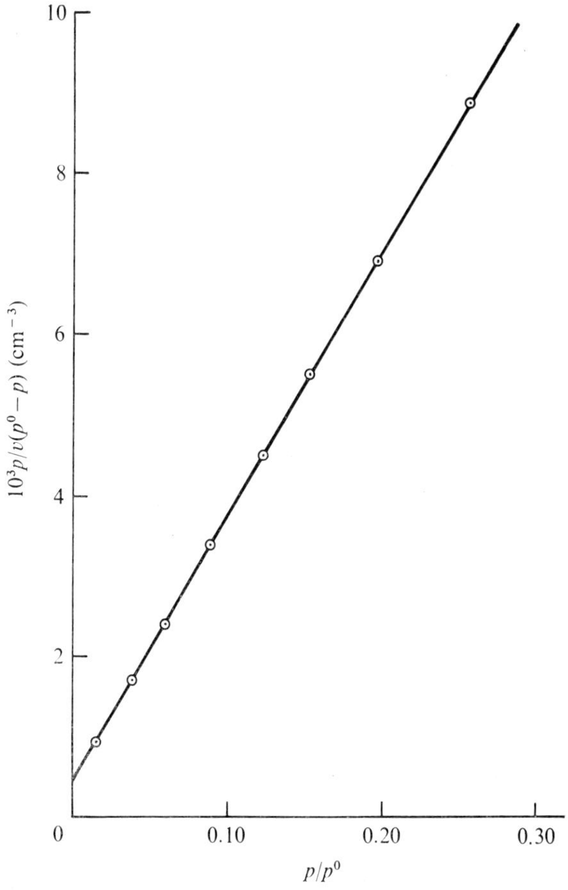

Fig. 5.8. Adsorption isotherm for N_2 on a fused copper catalyst at 90.1 K plotted according to (5.39). (After Brunauer, 1944.)

to the completion of a monolayer. Fig. 5.8 shows the plot according to (5.39) for nitrogen adsorption on a fused copper catalyst at 90.1 K. From v_m, and a knowledge of the cross-sectional area of the adsorbate molecule (which for N_2 is normally taken as 0.162 nm^2 molecule^{-1}), the surface area of the adsorbent may readily be calculated. Strictly, according to the model, the cross-sectional area of an adsorption site rather than that of the adsorbate molecule ought to be used, but the former is an unknown quantity. However, this does not appear to affect the usefulness of the BET equation for the assessment of surface areas.

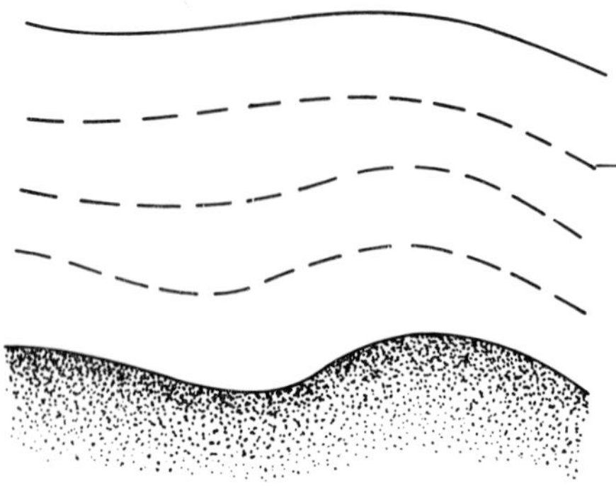

Fig. 5.9. Absorbed layer according to potential theory of adsorption.

5.13. The Polanyi potential theory of adsorption. Polanyi (1914) advanced a thermodynamic theory of gas adsorption. Because it proposes no detailed molecular model the theory is less open to criticism than is the BET treatment. The basic concept is that there is a force field surrounding a solid, which influences the adsorbate molecules. The forces are long-range and fall off with distance from the surface.

In the theory the adsorption potential, ϵ, is defined as the work done by adsorption forces in the transference of an adsorbate molecule from the bulk gas to a point in the surface phase. Since the force of attraction decreases with distance from the surface, ϵ also decreases, having a maximum value of ϵ_0 at the solid surface. It is possible to construct planes of equipotential around a solid as depicted in fig. 5.9.

The volumes of adsorbate enclosed between the solid surface and the equipotential planes defined by $\epsilon_1, \epsilon_2 \ldots$ are designated $\varphi_1, \varphi_2 \ldots$. As φ_i increases, ϵ_i decreases, and the interrelationship may be expressed in the general form

$$\epsilon_i = f(\varphi_i). \qquad \blacktriangleleft \textbf{(5.40)}$$

It is supposed that ϵ_i is independent of temperature. For a given system the curve represented by (5.40) is termed the 'characteristic curve'.

Consider the simple case of the adsorption of an ideal gas or vapour. Assume that the temperature is well below the critical temperature, so that the adsorbed film is liquid, and that the liquid is incompressible. Then ϵ_i is the work required to compress the ideal gas, at constant temperature, from the pressure, p_x, of the gas to the vapour pressure, p^0, of the liquid at that temperature. Thus, per mole,

$$\epsilon_i = RT \ln (p^0/p_x). \qquad (5.41)$$

The compression is accomplished by the force field of the solid. ϵ_i should not be confused with $\Delta_a \mu$ in (1.100), although $\epsilon_i = -\Delta_a \mu$ if p^0 is chosen as $p^\ominus$. The work of creating the interface between the liquid film and the gas is not here taken into account but this does not lead to serious errors.

The volume φ_i is given by

$$\varphi_i = m/\rho_l, \tag{5.42}$$

where m is the mass adsorbed at a pressure p_x and ρ_l is the density of the liquid adsorbate at the temperature of the adsorption. Corrections can be introduced into the treatment to allow for the compressibility of the adsorbed phase. Adsorption above the critical temperature, when the adsorbed phase is a compressed gas, has also been treated (Brunauer, 1944*a*).

The success of the theory lies in its ability to predict isotherms at any temperature from results obtained at only one temperature. From a single isotherm, values of ϵ_i and φ_i may be calculated in the simplest case, by the use of (5.41) and (5.42) so giving the 'characteristic curve' for the system. Isotherms for other temperatures may then be calculated by taking values of ϵ_i and of φ_i and finding m and p_x from (5.41) and (5.42), using the appropriate values of ρ_l and of p^0.

Brunauer quotes as an example of the use of the potential theory the adsorption of CO_2 on charcoal. The 'characteristic curve' is shown in fig. 5.10 where it can be seen that points obtained for widely differing temperatures both below and above the critical temperature, fall on the same curve. Fig. 5.11 shows the isotherms for the same system. The circles represent experimental data and the full lines the isotherms calculated from experimental results obtained at 273 K only. The agreement between experiment and theory is excellent.

More recently the potential theory has been used in the study of porosity of solid adsorbents (§5.21).

5.14. The Frenkel–Halsey–Hill slab theory. There have been several refined treatments of multilayer adsorption (see Young & Crowell, 1962*c*) which are beyond the scope of an introductory text. However, the so-called slab theory is a moderately simple example which yields a successful isotherm. The theory has been treated independently by Frenkel, Halsey and Hill but the exposition presented here is similar to that given by Hill (see Hill, 1952).

The theory bears a certain resemblance to the Polanyi theory (§5.13). For large surface coverages ($\theta > 2$) it is assumed that the detailed nature of the adsorbent surface is unimportant and that the environ-

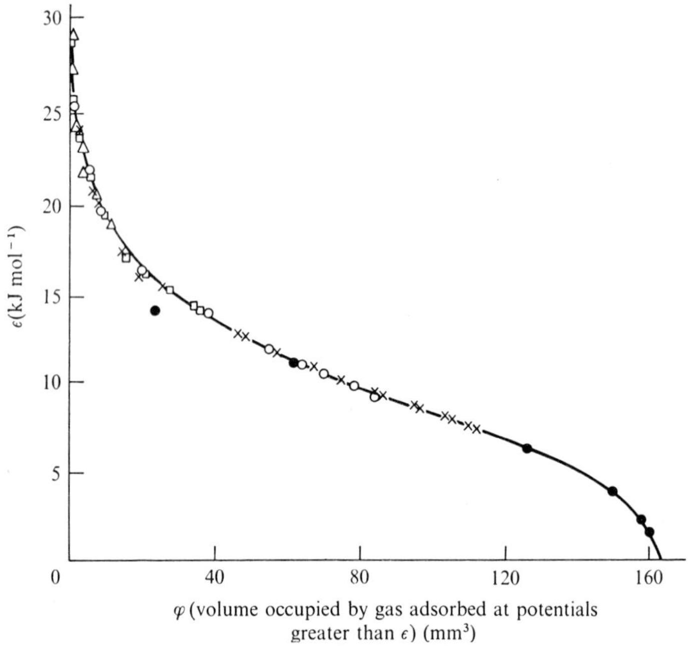

Fig. 5.10. Characteristic curve for CO_2 on charcoal.Data of Titoff (1910). ●, 196 K; ×, 273.1 K; ○, 303.1 K; □, 353.1 K; △, 424.6 K. (After Brunauer, 1944.)

ment of an adsorbate molecule (assumed to be non-polar) at the surface is similar to that in bulk liquid. The film is thus treated as a slab of liquid of uniform thickness in the potential field of the adsorbent. The adsorbent is assumed to be non-porous and non-polar, so that the adsorbate–adsorbent interactions are due solely to dispersion and repulsive forces.

Consider the adsorbed slab resting first on the solid (fig. 5.12(*a*)) and then on further liquid (fig. 5.12(*b*)). The slab has a uniform thickness h and the number of molecules per unit volume of the liquid is ρ_N^l. If the surface concentration of adsorbate is Γ^s, then $\rho_N^l = \Gamma^s/h$. It is assumed that the only significant difference between a molecule A in the slab in each of the two situations is one of potential energy.

For a molecule A at $z = h$, in fig. 12(*a*), there is an interaction Φ_x with the slab of liquid and a further interaction Φ_y with the solid. For A at $z = h$ in fig. 5.12(*b*) there is the same interaction Φ_x with the slab of liquid together with an interaction Φ_z with the underlying liquid.

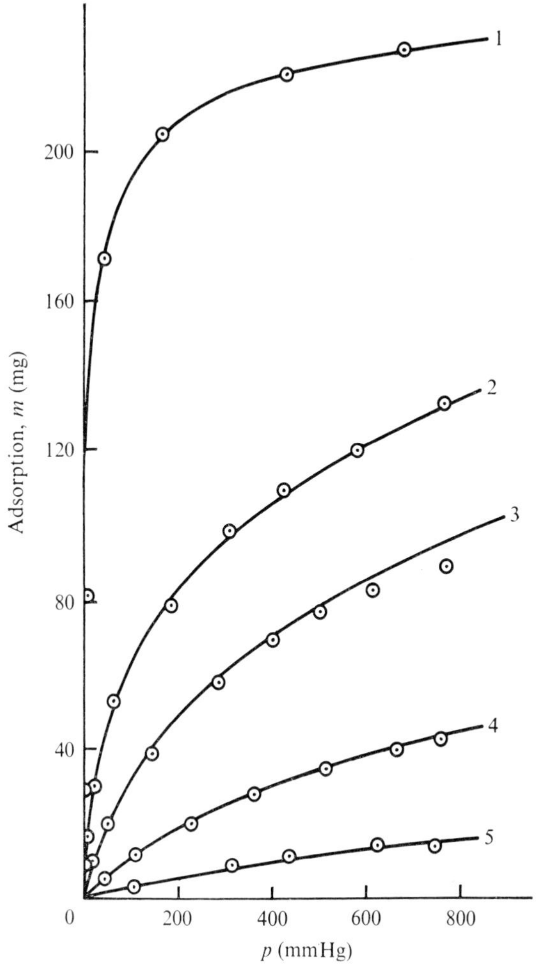

Fig. 5.11. Isotherms for the adsorption of CO_2 on charcoal. Points are experimental data of Titoff (1910), lines are calculated using potential theory from data at 273.1K. Curve 1, 196.6 K; 2, 273.1 K; 3, 301.1 K; 4, 353.1 K; 5, 424.6 K. (After Brunauer, 1944.)

The difference in potential energy of A in (*a*) and (*b*) is thus $(\Phi_y - \Phi_z)$. According to the model

$$\Phi_y - \Phi_z = \mu - \mu^l, \tag{5.43}$$

where μ is the chemical potential of a molecule in the adsorbed phase and μ^l that of a molecule in bulk liquid. In addition

$$\mu - \mu^l = kT \ln(p/p^0), \tag{5.44}$$

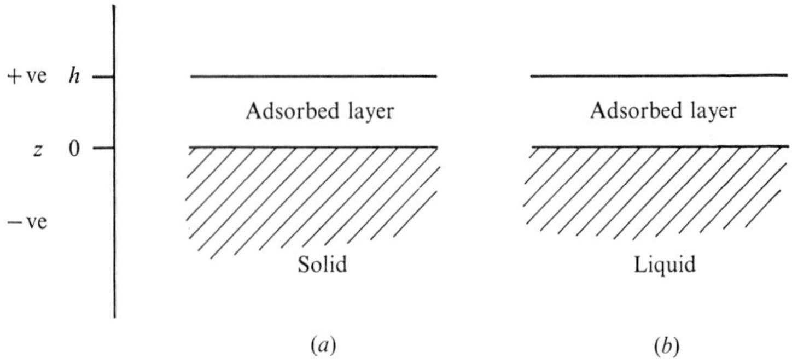

Fig. 5.12. See text.

where p is the equilibrium pressure of gas over the adsorbed slab in fig. 5.12(*a*).

From (5.8), if the small repulsive term in z^{-9} is ignored, Φ_y may be expressed

$$\Phi_y = -\pi\rho_N C/6h^3. \tag{5.45}$$

For the interaction of A with the liquid, the constant of attraction in (5.3) is written C_l and $\Phi_{\mathfrak{z}}$ is given by

$$\Phi_{\mathfrak{z}} = -\pi\rho_N^l C_l/6h^3. \tag{5.46}$$

From (5.43) to (5.46), noting that $h = \Gamma^s/\rho_N^l$,

$$\ln\frac{p}{p_0} = -\{[\pi(\rho_N^l)^3\rho_N C - \pi(\rho_N^l)^4 C_l]/6kT\}\frac{1}{(\Gamma^s)^3}$$

or

$$\ln\frac{p}{p^0} = -\frac{\beta}{(\Gamma^s)^3}, \tag{◀5.47}$$

where β is the constant in the curly brackets; (5.47) is the required isotherm.

Halsey (1948) obtained the similar but more general isotherm

$$\ln\frac{p}{p^0} = -\frac{b}{(\Gamma^s)^n}, \tag{5.48}$$

where b and n are constants. Halsey assumed that the interaction of a molecule varies with the inverse nth power of the distance from the surface, thereby obtaining $(\Gamma^s)^n$ rather than $(\Gamma^s)^3$.

The isotherm may be tested, and a value of n obtained by plotting $\ln\ln(p/p^0)$ against $\ln v$ (since v is proportional to Γ^s). It is found in practice that n is not generally equal to 3 as predicted by (5.47). Thus, for oxygen adsorbed on aluminium foil at -196 °C, $n = 2.7$ and for argon on rutile at -188 °C, $n = 2$. Values of n greater than 3 are also found. For example, for n-butanol adsorbed on anatase at 25 °C, $n = 6.0$. Halsey (1948) explains the variation in n in the following way. Large values of n indicate that the forces of attraction of the surface for the adsorbate fall off rapidly with distance, such as might occur for interactions of a specific nature. Smaller values of n result when the forces are more of the van der Waals type and extend over a greater distance. Thus the value of n typifies the nature of the adsorbent–adsorbate interactions.

5.15. Adsorption on porous solids: introduction. Many adsorbents are porous and, indeed, their utility frequently depends upon this property. Porosity enhances the surface area and provides spaces in which an adsorbate may condense.

The presence of pores introduces new problems into the treatment of the adsorption process. The amount of material which can be adsorbed on the surface of a pore is not unrestricted, as was assumed for the open surfaces considered in the BET treatment (§5.12). An additional type of interaction is also present. Consider a hypothetical case in which a pore has parallel walls separated by a distance of a few molecular diameters. Adsorption occurs on each of these walls and the gas–solid interactions decrease with distance from the surface. However, as the film thickness increases the films on opposite walls come closer to each other and experience mutual attractive forces. Adsorption in small pores is thus more preferred to that at plane surfaces. The pores are filled with liquid, therefore, at pressures less than p^0. This process is known as *capillary condensation.* For porous solids it is commonly found that the adsorption isotherm is not coincident with the desorption isotherm. This phenomenon is called *adsorption hysteresis.*

5.16. The classification of pores. It is convenient to have a classification of pores according to size. Pores with widths of less than about 1.5 nm are termed *micropores* whereas pores with widths greater than 100 nm are referred to as *macropores.* Pores of intermediate size are called *transitional pores* or *mesopores.*

Dubinin (1955) quotes values of pore sizes in porous carbonaceous adsorbents (active carbons). These adsorbents are of importance since they are widely used in industry. If the pores are considered to be cylindrical then the macropore radii are found to be of the order of 10^3 nm, and the radii of the mesopores are around 10 nm. Usually in active carbons the surface area originates mainly from micropores (between 400 and 900 $m^2 g^{-1}$) although the volume of the micropores is only between 0.15 and 0.5 $cm^3 g^{-1}$. This can be compared with the contribution of macropores in the same specimens where, although their volume is very similar, their area is only of the order of 1 $m^2 g^{-1}$. A brief discussion of some of the methods employed in the study of porosity will be given later. First, however, an outline of some of the theories of adsorption on porous solids is presented.

5.17. Modification of the BET theory. It will be recalled that in the BET treatment (§5.12) no restriction was placed on the number of layers of adsorbate which could be adsorbed. However, in pores the adsorption is clearly restricted. If it is supposed that adsorption takes place on two parallel plane walls of a capillary, and that n is the maximum number of layers of adsorbate which can be accommodated on each of the walls, the BET equation becomes (Brunauer, 1944*b*)

$$\frac{v}{v_m} = \frac{cx}{(1-x)} \frac{[1-(n+1)x^n+nx^{n+1}]}{[1+(c-1)x-cx^{n+1}]}. \tag{5.49}$$

The symbols have the same significance as in (5.37). When $n = 1$, (5.49) reduces to the Langmuir equation and, when $n = \infty$, (5.37) results.

The effect of having a finite value of n in (5.49) is to reduce the calculated adsorption, as compared to that predicted by (5.37). For $x > 0.35$ this usually results in a better fit of the experimental data. An example, taken from Brunauer is shown in fig. 5.13.

The 'n-layer treatment' mentioned above takes no account of the forces of capillary condensation and (5.49) does not therefore describe isotherms of types IV and V (fig. 5.3). A modification of the BET equation has, however, been derived (see Brunauer, 1944*c*) in which it is assumed that the last layer adsorbed in the capillary yields a greater heat than adsorption in the other layers (excluding the first layer). It is argued that the introduction of the last layer of molecules destroys the two liquid surfaces, and liberates the heat of destruction of the surface. The equations derived have four adjustable parameters and are rather complex, for which reasons they are seldom used.

5.18. The Kelvin equation. Adsorption on porous solids involves the condensation of adsorbate in fine capillaries and, in consequence, a form

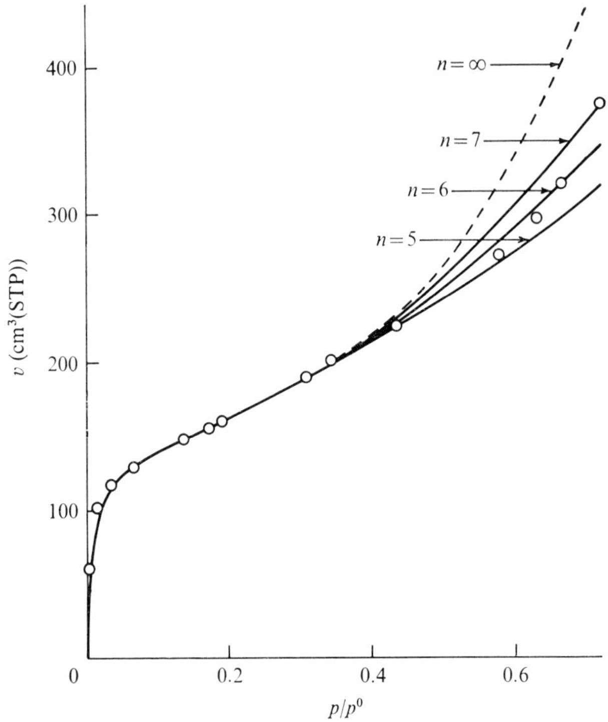

Fig. 5.13. Adsorption isotherm for nitrogen on an iron catalyst at 77.3 K. Points are experimental values, and the lines are calculated for various values of n using (5.49). v_m and c were obtained using (5.39) for p/p^0 up to 0.35. (After Brunauer, 1944.)

of the Kelvin equation (3.10) is very useful. The equation expresses the relationship between the vapour pressure p, over a curved surface such as that encountered in a capillary filled with condensed adsorbate, and the radius of curvature of the surface. For a concave spherically curved surface in a cylindrical capillary of radius r_t the equation is

$$\ln\frac{p}{p^0} = -\frac{2V_m\gamma\cos\theta}{r_t RT}. \tag{3.10}$$

V_m is the molar volume of the liquid and γ the surface tension. Equation (3.10) indicates that for values of the contact angle θ between 0° and 90°, $p < p^0$, i.e. liquid condenses in the capillary at a pressure less than the saturated vapour pressure of the adsorbate at the temperature T. Since (3.10) is a thermodynamic relationship it is valid only for condensation

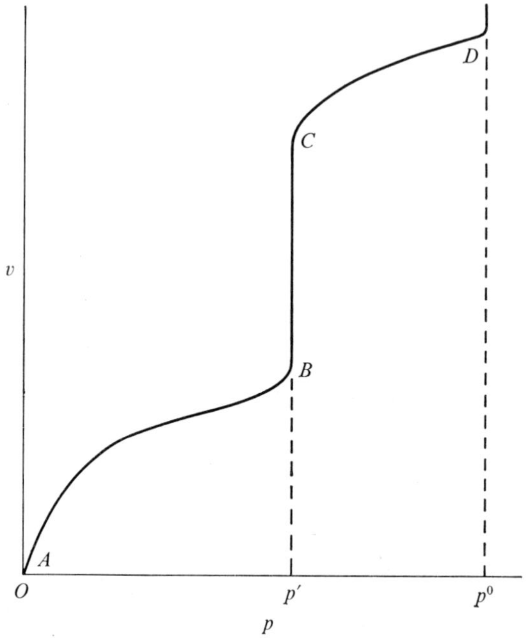

Fig. 5.14. Isotherm for adsorption on a solid having cylindrical pores all of radius r.

in pores whose radii are sufficiently large for the definition of γ and the concept of a meniscus to be thermodynamically meaningful. For this reason the equation is probably not applicable to liquid in micropores. At the other extreme, for macropores, r_t is sufficiently large for p and p^0 to have very similar values. The equation is therefore most useful in the treatment of condensation in mesopores.

Consider adsorption on a porous solid for which the pores may be assumed to be cylinders all of radius r_t, where r_t is sufficiently large for the Kelvin equation to be applicable. The adsorption at low pressures, represented by AB in fig. 5.14, can be described by the simple BET equation, or possibly by a suitable monolayer equation. If the adsorbent contains micropores it has been suggested that these are filled in the region AB. At a pressure p' condensation takes place in the capillaries. The radius r_t of the capillaries is related to p' by (3.10). At p' the isotherm rises vertically along BC, and at C the capillaries are full to the meniscus a depicted in fig. 5.15. The curvature of a is the same as the curvature of the meniscus in a partially filled capillary. As further

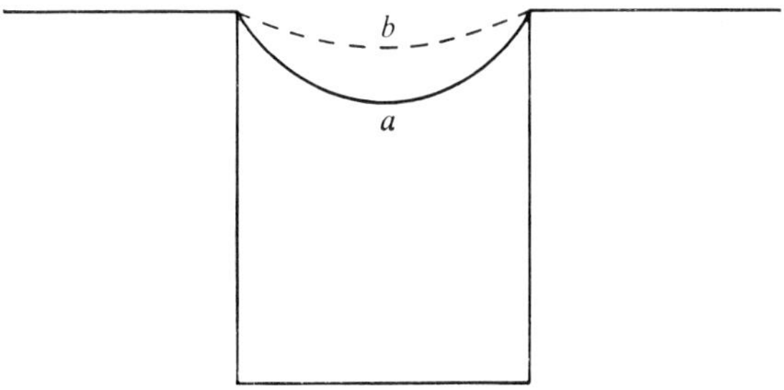

Fig. 5.15. Change in the curvature of a meniscus with increasing pressure.

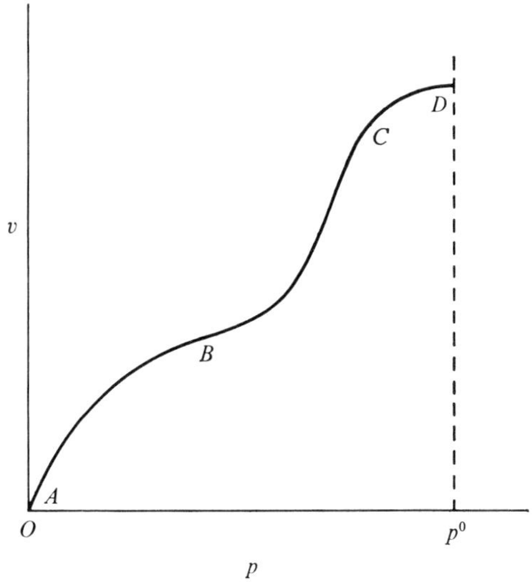

Fig. 5.16. An isotherm for adsorption on real porous solids.

condensation takes place at pressures greater than p' the meniscus becomes less curved, as represented by b and at $p = p^0$ the meniscus is planar. The section CD in fig. 5.14 includes the filling of the pore from meniscus a to the plane surface. In practice there is a distribution of pore sizes in solids and so the section BC for a real solid might be expected to have a sigmoid shape, as in fig. 5.16, and an isotherm of type IV is found.

5.19 Adsorption hysteresis. It is commonly found for porous adsorbents that the adsorption and the desorption branches of an isotherm are not

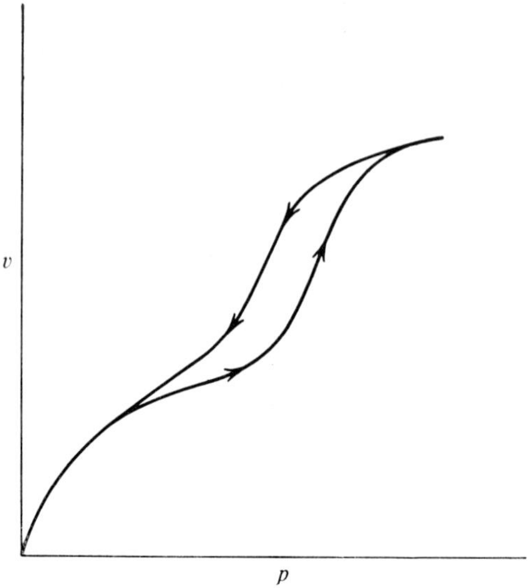

Fig. 5.17. Adsorption hysteresis involving an isotherm of type IV.

coincident over the whole pressure range. This phenomenon is termed hysteresis and an example, involving a type IV isotherm, is shown in fig. 5.17. The arrows indicate the direction of the pressure change. Hysteresis usually occurs over the region of the isotherm in which mesopores are being filled or emptied. The so-called hysteresis 'loops' are usually reproducible, and the desorption branch is always displaced to lower pressures.

Zsigmondy (1911) attempted to explain hysteresis in terms of the hysteresis of contact angle. It was supposed that during adsorption $\theta > 0$ as a result of the presence of adsorbed impurities on the walls of the pores. After adsorption it was assumed that the impurities had been displaced, and desorption occurred from a clean surface for which θ was zero. For such a mechanism to be applicable the hysteresis loop must disappear after the first adsorption–desorption cycle, and therefore reproducible hysteresis cannot be explained in this way.

Several other explanations for hysteresis have been proposed including the well-known 'ink bottle' theory. This latter is applicable when the pore entrances are narrower than the interior of the pores, as shown in fig. 5.18. During adsorption, the pore is filled when the pressure, p_a, corresponding to the condensation in the widest part of the pore, is

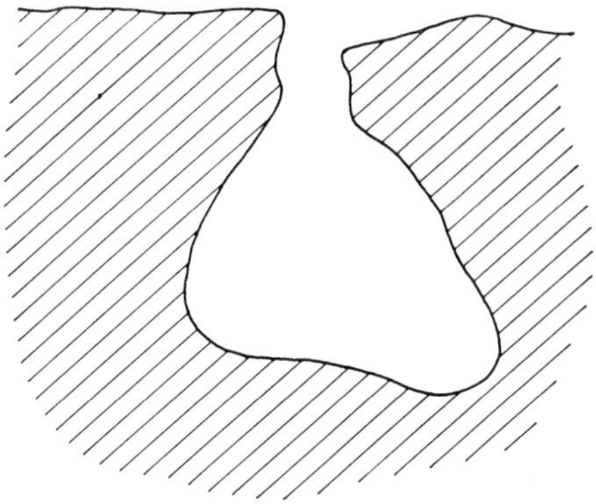

Fig. 5.18. Pore with narrow entrance and wide interior.

attained. During desorption however, the pore does not empty until the pressure p_d is reached where p_d corresponds to evaporation from a pore of radius equal to the neck radius. Clearly $p_a > p_d$, so that the desorption branch is shifted to lower pressures. Such a mechanism has been suggested by Kraemer (see Taylor, 1931) and by McBain (1935).

It is possible, by examination of the desorption branch of an isotherm, and by use of the Kelvin equation, to calculate a pore size distribution for an adsorbent. This problem is discussed by Gregg (1965*b*).

5.20. Pore size distribution using mercury porosimetry. If a non-wetting liquid ($\theta > 90°$) is brought into contact with a porous solid an external pressure must be applied to effect the entry of the liquid into the pores (§3.2). The Laplace equation (3.3), for a spherically curved meniscus in a cylindrical capillary of radius r_t is

$$\Delta p = \frac{2\gamma \cos \theta}{r_t}, \tag{3.3}$$

where Δp is the pressure drop over the curved surface.

The principle of mercury porosimetry can be demonstrated in the following way. Consider a single cylindrical pore, which has been outgassed, in contact with the non-wetting liquid mercury. If the mercury is forced into the capillary a convex surface is created as depicted in fig. 5.19. Since the pressure at *A* is zero the minimum pressure required for the mercury to enter the pore is equal to the pressure drop

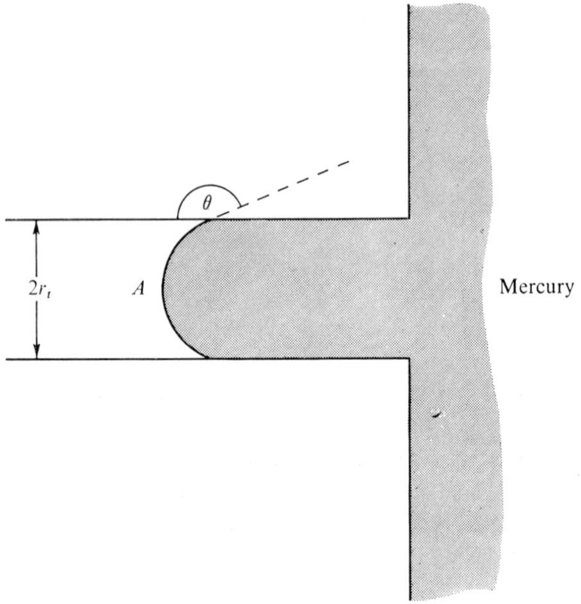

Fig. 5.19. Mercury penetrating a cylindrical pore.

over the curved surface, i.e. $|2\gamma \cos \theta / r_t|$. Thus, from a knowledge of the pressure at which the mercury enters the pore, and of θ and γ, the radius of the pore can be calculated. Values of Δp necessary to force mercury into pores of various diameters are given in table 5.1. The surface tension of mercury and the contact angle have been taken as 480 mN m^{-1} and 140° respectively.

Ritter & Drake (1945) have studied the pore structure of several adsorbents using mercury porosimetry, and describe a suitable apparatus. The essential part of the porosimeter is a glass dilatometer, in the bulb of which is placed a sample (between 1 and 20 g) of adsorbent. The solid sample is immersed in mercury, which extends part of the way up the calibrated glass capillary tube. Pressure is exerted on the mercury (using N_2 or, for higher pressures, oil) so forcing it into the pores of the solid. As the penetration takes place the level of the mercury in the capillary falls, and the volume of the pores filled with mercury at any pressure may be calculated. Since the dilatometer is enclosed in a metal bomb an indirect method for the determination of the height of the mercury is necessary. This may be achieved by means of a platinum

TABLE 5.1. *Variation of applied pressure p with pore diameter, from* (3.3)

Pressure		Pore	Pressure		Pore
p(MN m^{-2})	p (atmosphere)	diameter (nm)	p(MN m^{-2})	p (atmosphere)	diameter (nm)
0.172	1.70	8550	40	395	37
10	98.7	147	50	493	29
20	197	74	60	592	25
30	296	49	70	691	21

wire placed in the capillary and comprising one arm of a resistance bridge. Only the part of the wire outside the mercury contributes significantly to the resistance.

Let the total pore volume of a sample be V_0, and the volume of pores with radii less than r be V. It is supposed that the volume determined on the porosimeter at pressure p is the volume of pores with radii $\geqslant r$. This volume is therefore $(V_0 - V)$. Fig. 5.20 shows results obtained by Ritter and Drake for two adsorbents, an activated clay and fullers' earth (a hydrated aluminium silicate used industrially as an adsorbent). The pressure is related to the radius by (3.3) and the corresponding pore diameters have been included on the abscissa. The volume of pores with diameters greater than and equal to a certain value may be read directly from the graph.

The pore size distribution is determined in the following manner. Suppose the volume of pores having radii between r and $r+\mathrm{d}r$ is $\mathrm{d}V$. Then a distribution function $D(r)$, for the pore size is given by

$$D(r) = \mathrm{d}V/\mathrm{d}r. \tag{5.50}$$

From (3.3), assuming that θ and γ are constant

$$p\,\mathrm{d}r + r\,\mathrm{d}p = 0, \tag{5.51}$$

where p rather than Δp is written for the applied pressure and r_t is written r. From (5.50) and (5.51)

$$D(r) = -\frac{p}{r}\frac{\mathrm{d}V}{\mathrm{d}p}. \tag{5.52}$$

In addition

$$\frac{\mathrm{d}(V_0 - V)}{\mathrm{d}p} = -\frac{\mathrm{d}V}{\mathrm{d}p} \tag{5.53}$$

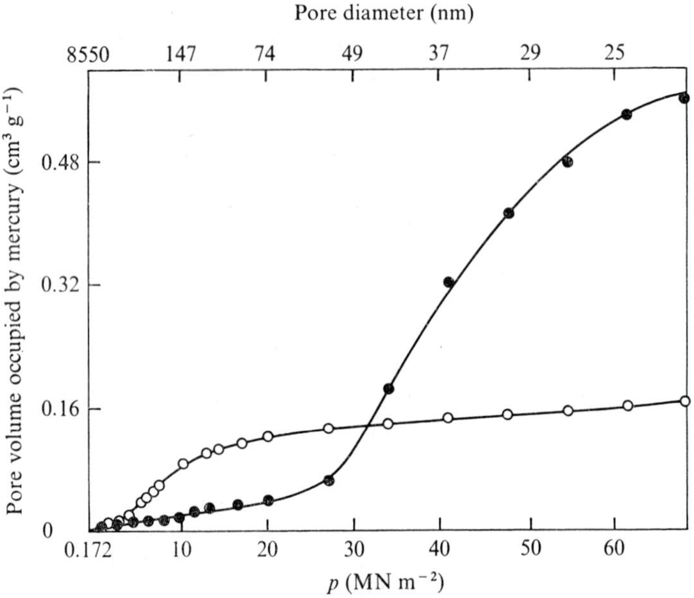

Fig. 5.20. Pore volumes of fullers' earth (●) and activated clay (○). (After Drake & Ritter, 1945.)

so that from (5.52) and (5.53)

$$D(r) = \frac{p}{r}\frac{\mathrm{d}(V_0 - V)}{\mathrm{d}p}. \tag{5.54}$$

The pore size distribution curve is a plot of $D(r)$ against r, examples of which are shown in fig. 5.21. The maxima in the curves correspond to the radii of the pores present in greatest abundance. It should be remembered that the above procedures are based on the assumption of cylindrical pores and therefore only qualitative indications concerning the pore structure are obtained.

5.21. The use of the potential theory for microporous solids. This application, which has been largely developed by Russian workers, is discussed by Dubinin (1960).

The potential fields of opposite walls in a micropore overlap, and this causes the condensation of adsorbate at very low relative pressures. Under the circumstances it is proposed that the characteristic curve (see §5.13) may be written explicitly as

$$\varphi = \varphi_0 \exp[-k\epsilon^2], \tag{5.55}$$

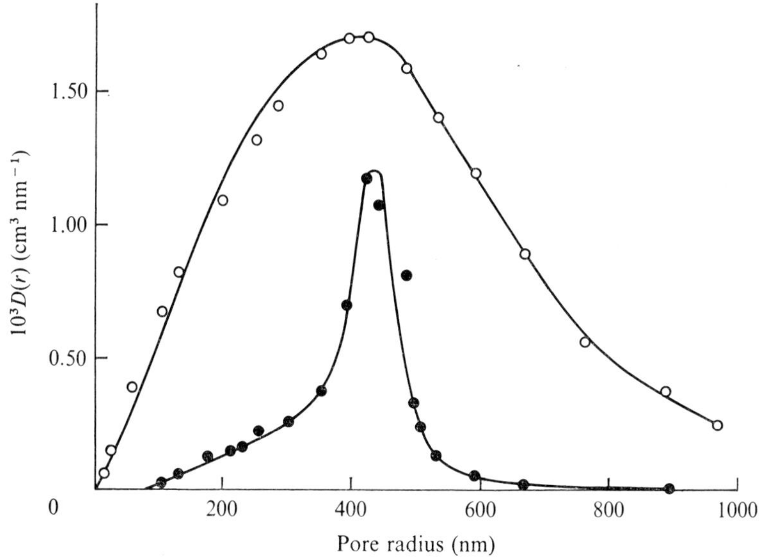

Fig. 5.21. Distribution functions for diatomaceous earth (○) and fritted glass (●). (After Ritter & Drake, 1945.)

where φ is the volume adsorbed (in cm³ liquid) and represents the volume of micropores filled at any pressure p. The term φ_0 is the limiting adsorption, i.e. the total micropore volume, ϵ is the adsorption potential given by (5.41), and k is a constant. The exponent 2 arises from the overlap of the potential fields. Equation (5.55) expressed in logarithmic form becomes

$$\ln \varphi = \ln \varphi_0 - k\epsilon^2. \tag{5.56}$$

Thus, in the low pressure region of an isotherm, where micropore filling is taking place, a plot of $\ln \varphi$ against ϵ^2 should be linear, with an intercept at $\epsilon^2 = 0$ of $\ln \varphi_0$. Since, according to the theory, φ_0 is a property of the solid, the intercept should be independent of the adsorbate, provided that the molecular size of the latter is not too great to allow entry into the micropores.

Examples of linear plots according to (5.56) for five adsorbates on two microporous carbon samples are given in fig. 5.22. The experiments were carried out over a range of temperatures from -196 °C (for N_2) to 50 °C (for C_2F_4 and C_3F_6). Values of φ/φ_0 range from 0.06 to 0.94.

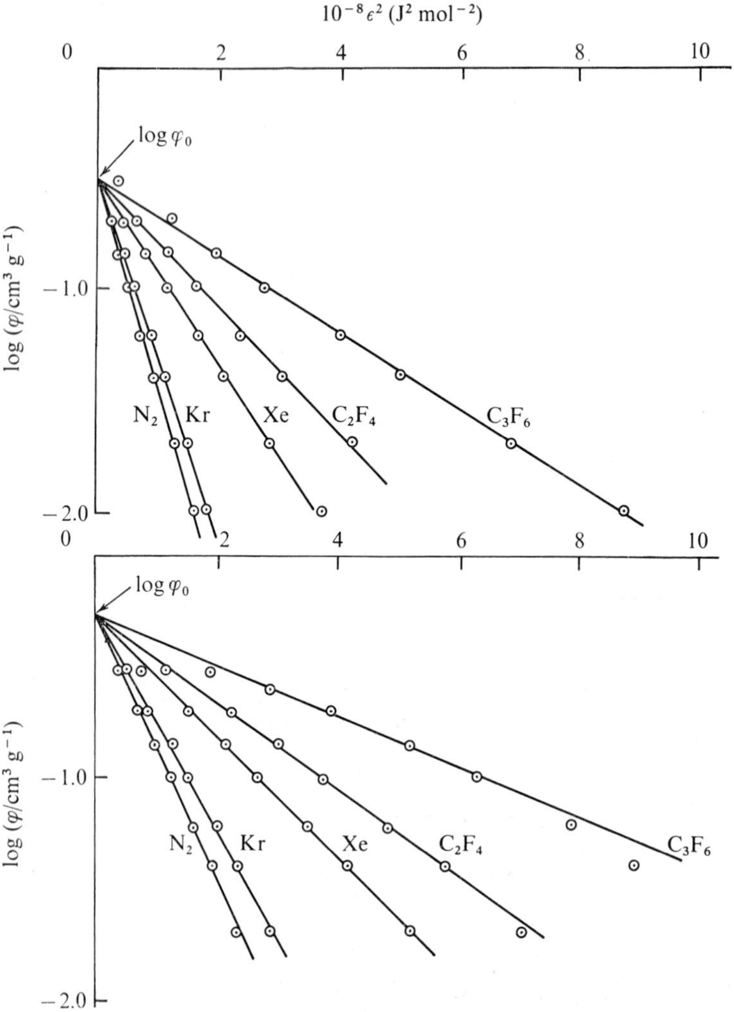

Fig. 5.22. Plots according to (5.56) for various gases adsorbed on two different active carbons. (After Dubinin, 1960.)

5.22. Thermodynamics of adsorption of gases on solids. This subject is both extensive and complex, as may be appreciated by reference to Hill (1952), Young & Crowell (1962*d*) and Ross & Olivier (1964). The object of this section is mainly to introduce some of the useful thermodynamic parameters which can be obtained from adsorption isotherms.

A more complete account along the same lines is given by Young and Crowell.

It will be supposed that the solid simply provides a potential field which causes adsorption, and that the thermodynamic properties of the solid remain unchanged by the adsorption. The assumed inert nature of the solid appears to be physically realistic in many instances. In view of this, equations analogous to those derived for the whole interface in §1.7 will here be used for the adsorbed layer only. It is only necessary to replace γ by $-\pi$, where π, the surface pressure, is analogous to p for a one component three-dimensional phase.† Thus, for the adsorption of a single gas, (1.8) becomes

$$\mathrm{d}U^s = T\mathrm{d}S^s - p\mathrm{d}V^s - \pi\mathrm{d}\mathcal{A} + \mu\mathrm{d}n^s \qquad (5.57)$$

and (1.19) may be expressed

$$\mathrm{d}\mathcal{G}^s = -S^s\mathrm{d}T + V^s\mathrm{d}p - \pi\mathrm{d}\mathcal{A} + \mu\mathrm{d}n^s. \qquad (5.58)$$

$\mathcal{A}$, the surface area of the adsorbent, is proportional to the number of moles n_A of adsorbent in the system, and n^s is the number of moles of adsorbed gas. The hydrostatic pressure, p, is, under normal experimental conditions, equal to the pressure of the gas over the solid. This would not be the case if for example a second non-adsorbable gas were present, but for this discussion the hydrostatic and adsorbate pressures will be supposed to be equivalent.

Entropies of adsorption. It follows from (5.58) that

$$\mathrm{d}\mu = -\bar{\mathrm{s}}^s\mathrm{d}T + \bar{\mathrm{v}}^s\mathrm{d}p - \left(\frac{\partial\pi}{\partial n^s}\right)_{p,T,a}\mathrm{d}\mathcal{A} + \left(\frac{\partial\mu}{\partial n^s}\right)_{p,T,a}\mathrm{d}n^s, \qquad (5.59)$$

where $\bar{\mathrm{s}}^s$ and $\bar{\mathrm{v}}^s$ are differential quantities defined by

$$\bar{\mathrm{s}}^s = \left(\frac{\partial S^s}{\partial n^s}\right)_{p,T,a} \quad \text{and} \quad \bar{\mathrm{v}}^s = \left(\frac{\partial V^s}{\partial n^s}\right)_{p,T,a}. \qquad (5.60)$$

It is possible to obtain from experimental isotherms measured over a range of temperature, the variation of p with T for a constant amount, n^s, of adsorbate at the surface and constant $\mathcal{A}$ (i.e. constant Γ^s). For $\mathrm{d}n^s = 0$ and $\mathrm{d}\mathcal{A} = 0$, (5.59) becomes

$$\mathrm{d}\mu = -\bar{\mathrm{s}}^s\mathrm{d}T + \bar{\mathrm{v}}^s\mathrm{d}p. \qquad (5.61)$$

† See appendix to this chapter.

For the gas, which is in equilibrium with the adsorbed phase,

$$\mathrm{d}\mu = -\mathrm{s}^g\,\mathrm{d}T + \mathrm{v}^g\,\mathrm{d}p, \tag{5.62}$$

where s^g and v^g are the molar quantities for the unadsorbed gas. From (5.61) and (5.62) it follows that

$$\left(\frac{\partial p}{\partial T}\right)_{n^s,a} = \frac{\mathrm{s}^g - \bar{\mathrm{s}}^s}{\mathrm{v}^g - \bar{\mathrm{v}}^s}. \tag{5.63}$$

If $\bar{\mathrm{v}}^s$ is neglected in comparison with v^g, and if the gas is assumed ideal then (5.63) may be written

$$\left(\frac{\partial \ln p}{\partial T}\right)_{n^s,a} = \frac{\mathrm{s}^g - \bar{\mathrm{s}}^s}{RT}. \tag{◀5.64}$$

Thus, the quantity $\mathrm{s}^g - \bar{\mathrm{s}}^s$ may be obtained from adsorption isosteres, i.e. plots of p against T at constant n^s.

s^g, the molar entropy of the gas, refers to a particular temperature and pressure p. It is often convenient to obtain the difference between $\bar{\mathrm{s}}^s$ and the entropy of the gas in a reference state (at the same temperature). If unit pressure is chosen and $\mathrm{s}^{g\prime}$ is the molar entropy of the gas in this state, then

$$\mathrm{s}^{g\prime} - \bar{\mathrm{s}}^s = (\mathrm{s}^g - \bar{\mathrm{s}}^s) + R\ln p. \tag{5.65}$$

If the reference state of pure liquid at temperature T is chosen, where the molar entropy is denoted $\mathrm{s}^{l\prime}$, then

$$\mathrm{s}^{l\prime} - \bar{\mathrm{s}}^s = (\mathrm{s}^g - \bar{\mathrm{s}}^s) + R\ln(p/p^0) - \Delta_e H/T, \tag{5.66}$$

where $\Delta_e H$ is the molar heat of vaporisation of the adsorbate and p^0 the saturated vapour pressure, both at temperature T. Whatever standard states are chosen, however, it will be noted that (5.64) yields the differential molar entropy of the adsorbed gas.

An expression which may be used to obtain the molar entropy s^s of the adsorbed gas can be derived as follows. The expression for $\mathrm{d}G^s$ is (see 1.22)

$$\mathrm{d}G^s = -S^s\mathrm{d}T + V^s\mathrm{d}p + \mathcal{A}\,\mathrm{d}\pi + \mu\,\mathrm{d}n^s. \tag{5.67}$$

In addition, from (1.24),

$$\mathrm{d}G^s = \mu\,\mathrm{d}n^s + n^s\mathrm{d}\mu. \tag{5.68}$$

Combination of (5.67) and (5.68) leads to

$$\mathrm{d}\mu = -\mathrm{s}^s\mathrm{d}T + \mathrm{v}^s\mathrm{d}p + \frac{\mathcal{A}}{n^s}\mathrm{d}\pi, \tag{5.69}$$

where $s^s = S^s/n^s$ and $v^s = V^s/n^s$. From (5.62) and (5.69), holding π constant ($d\pi = 0$),

$$\left(\frac{\partial p}{\partial T}\right)_\pi = \frac{s^g - s^s}{v^g - v^s} \tag{5.70}$$

or, making the same assumptions as for (5.64),

$$\left(\frac{\partial \ln p}{\partial T}\right)_\pi = \frac{s^g - s^s}{RT}. \qquad \blacktriangleleft\textbf{(5.71)}$$

To obtain s^s from (5.71) it is necessary to calculate π from adsorption data; this may be accomplished by the use of (5.13). It is interesting to note that (5.13) may be obtained from (5.62) and (5.69). Thus, for constant T, the combination of these two equations yields

$$\left(\frac{\partial \pi}{\partial p}\right)_T = \Gamma^s(v^g - v^s), \tag{5.72}$$

where $\Gamma^s = n^s/\mathcal{A}$. By noting that $v^g \gg v^s$, and assuming ideal gas behaviour it follows from (5.72) that

$$\pi = RT\int_{p=0}^{p} \Gamma^s \mathrm{d}\ln p. \tag{5.13}$$

The difference between s^s and the molar entropy of the adsorbate at a standard pressure of unity or in the pure liquid state, can be obtained from equations analogous to (5.65) and (5.66).

Heats of adsorption. There are various heats of adsorption (Everett, 1950). The heat obtained in a particular experiment depends on the conditions under which adsorption occurs. In general, if the heat given out under isothermal conditions for the adsorption of n^s moles of adsorbate on an initially clean surface is $-Q$, then a differential heat of adsorption q is defined as

$$q = -\mathrm{d}Q/\mathrm{d}n^s. \tag{5.73}$$

Two differential heats of adsorption may be obtained from (5.64) and (5.71), i.e.

$$-\Delta_a\overline{\mathcal{H}} = q_{st} = T(s^g - \bar{s}^s) \tag{5.74}$$

and

$$-\Delta_a \mathrm{H} = T(s^g - s^s) \tag{5.75}$$

so that

$$\left(\frac{\partial \ln p}{\partial T}\right)_{\Gamma^s} = \frac{q_{st}}{RT^2} = -\frac{\Delta_a\overline{\mathcal{H}}}{RT^2} \qquad \blacktriangleleft\textbf{(5.76)}$$

and
$$\left(\frac{\partial \ln p}{\partial T}\right)_\pi = -\frac{\Delta_a \mathrm{H}}{RT^2}. \quad \blacktriangleleft(5.77)$$

The quantity q_{st} is termed the *isosteric* heat of adsorption and, as can be seen from (5.73), is conventionally taken as positive for the exothermic adsorption process. The quantity $\Delta_a \mathrm{H}$ is often called the *equilibrium* heat of adsorption.†

It can be seen from (5.74) and (5.75) that

$$\Delta_a \mathrm{H} = \Delta_a \overline{\mathscr{H}} + T(\mathrm{S}^s - \bar{\mathrm{S}}^s). \quad (5.78)$$

In addition, for constant n^s, $\mathscr{A}$ and p, it follows from (5.61) and (5.69) that

$$-\bar{\mathrm{S}}^s \mathrm{d}T = -\mathrm{S}^s \mathrm{d}T + \frac{\mathscr{A}}{n^s} . \mathrm{d}\pi_{p,\mathscr{A}, n^s}. \quad (5.79)$$

Noting that $\mathscr{A}/n^s = 1/\Gamma^s$, (5.79) leads to

$$\mathrm{S}^s - \bar{\mathrm{S}}^s = \frac{1}{\Gamma^s}\left(\frac{\partial \pi}{\partial T}\right)_{p, \Gamma^s}. \quad (5.80)$$

Combination of (5.80) and (5.78) gives the relationship between $\Delta_a \mathrm{H}$ and $\Delta_a \overline{\mathscr{H}}$ as

$$\Delta_a \mathrm{H} = \Delta_a \overline{\mathscr{H}} + \frac{T}{\Gamma^s}\left(\frac{\partial \pi}{\partial T}\right)_{p, \Gamma^s}. \quad \blacktriangleleft(5.81)$$

The two heats of adsorption may be expressed as

$$\Delta_a \mathrm{H} = \mathrm{H}^s - \mathrm{H}^g \quad (5.82)$$

and
$$\Delta_a \overline{\mathscr{H}} = \overline{\mathscr{H}}^s - \mathrm{H}^g, \quad (5.83)$$

where H^g is the molar heat content of the gas. It can be shown that

$$n^s \mathrm{H}^s = U^s + pV^s + \pi \mathscr{A},$$

and that
$$\overline{\mathscr{H}}^s = (\partial \mathscr{H}^s / \partial n^s)_{T, p, \mathscr{A}},$$

† The different significance of the two entropies S^s and $\bar{\mathrm{S}}^s$ can be understood in the following terms. Various differential entropies $(\partial S^s/\partial n^s)$ may be defined depending on which properties are held constant during the addition of adsorbate to the surface. If all the intensive properties (T, p, π) are held constant then in a one component system the differential entropy is identical to the molar entropy, i.e. $(\partial S^s/\partial n^s)_{T,p,\pi} = S^s/n^s = \mathrm{S}^s$. In such a process, since the intensive properties are constant, the same equilibrium position is maintained throughout as, for example, in fusion and vaporisation. This is why $\Delta_a \mathrm{H}$, which is related to S^s by (5.75), is termed the equilibrium heat of adsorption. If, however, for the addition of adsorbate to the surface, $\mathscr{A}$ rather than π is held constant, the differential entropy $\bar{\mathrm{S}}^s$ (5.60) is obtained.

where $$\mathscr{H}^s = U^s + pV^s.$$

The heat contents are related to the Gibbs functions (see §1.7) by

$$\mathscr{G}^s = \mathscr{H}^s - TS^s \quad \text{and} \quad G^s = H^s - TS^s.$$

A further heat frequently encountered is the so-called *differential* heat of adsorption, q_d. The significance of q_d may be shown as follows. $\mathrm{d}Q$ at constant $\mathscr{A}$ is given by (see 1.1)

$$\mathrm{d}Q = \mathrm{d}U + p\mathrm{d}V. \tag{5.84}$$

If, for isothermal adsorption, V^g and V^s are held constant

$$\mathrm{d}Q_{V^g, V^s, a} = \mathrm{d}U^g + \mathrm{d}U^s. \tag{5.85}$$

Since $U^g = n^g\mathrm{U}^g$ and $\mathrm{d}n^g = -\mathrm{d}n^s$, it follows from (5.85) and (5.73) that

$$q_{V^g, V^s, a} = \mathrm{U}^g - \overline{\mathrm{U}}^s = -\Delta_a\overline{\mathrm{U}} = q_d \tag{5.86}$$

and for an ideal gas, to a good approximation

$$q_d = q_{st} - RT. \tag{◄ 5.87}$$

The quantity $-\Delta_a\mathrm{H}$ is analogous to the heat of vaporisation of a liquid at its saturated vapour pressure and $(\mathrm{S}^g - \mathrm{S}^s)$, (5.71), is analogous to the molar entropy of vaporisation. $\Delta_a\mathrm{H}$ is not measured calorimetrically but is obtained from adsorption data by the use of (5.77). This procedure requires that values of π be calculated from adsorption isotherms with the aid of the Gibbs equation (5.13). q_d can be obtained calorimetrically† and q_{st} may then be calculated from (5.87). Alternatively q_{st} can be obtained from adsorption isotherms obtained at various temperatures, by use of (5.76).

Comparison of experimental and theoretical entropies of adsorption. It has been stressed (§1.9 and §5.8) that agreement between experimental adsorption data obtained at a single temperature, and a theoretical isotherm, is insufficient evidence of the validity of the model. If the adsorption is studied as a function of temperature, however, it is possible to calculate entropies and heats of adsorption which can then be compared with values predicted from a model. Kemball (see Kemball, 1950) did much of the early work on the comparison of experimental

† The calorimetric heat is not exactly equal to $-\Delta_a\overline{\mathrm{U}}$ since although V^g and $\mathscr{A}$ can be held constant for the adsorption (see 5.86), V^s cannot. However, the difference between the two quantities is usually negligible.

and theoretical entropies of adsorption. It should be appreciated, nonetheless, that even this procedure may not lead to definitive conclusions as to the nature of an adsorbed film.

De Boer & Kruyer (1952) have compared the experimental differential entropies of adsorption for a number of systems, with the theoretical values calculated on the assumption of both a localised and a non-localised monolayer. For the localised film it was assumed that the internal rotations and vibrations of the adsorbate are retained on adsorption and that vibrations parallel to and perpendicular to the surface do not contribute to the entropy. The differential entropy change resulting from the adsorption of a mole of gas to form such a localised film is therefore

$$\bar{s}^s_{config} - s^g_{tr}, \tag{5.88}$$

where s^g_{tr} is the molar translational entropy of the gas and $\bar{s}^s_{config}$ the differential molar configurational entropy of the adsorbed gas. In the model for non-localised adsorption it is supposed that the entropy of adsorption results only from the loss of translational entropy, and is therefore

$$\bar{s}^s_{tr} - s^g_{tr}, \tag{5.89}$$

where $\bar{s}^s_{tr}$ is the differential molar translational entropy of a two-dimensional ideal gas. Values of $\bar{s}^s_{config}$, $\bar{s}^s_{tr}$ and s^g_{tr} can all be calculated theoretically.

For the sake of comparison of experimental and theoretical values of the differential entropies of adsorption it is convenient to choose standard states for the gas and the adsorbed film (see §1.14) and then to compare the standard entropies. A standard pressure of one atmosphere is used for the gas. Different standard states were chosen by de Boer and Kruyer for the localised and the non-localised films as will be seen.

The standard value $s^{g,\ominus}_{tr}$ for an adsorbate, assumed ideal, is calculated for the appropriate temperature using the Sackur–Tetrode equation. The differential molar entropy ($\bar{s}^s_{config}$) can be obtained by differentiating the entropy, with respect to N, of the system of N molecules distributed over N_s sites (see §1.12 where a similar calculation is made to obtain an expression for μ). It is readily shown that

$$\bar{s}^s_{config} = -R \ln \frac{\theta}{1-\theta}. \tag{5.90}$$

The standard state for the localised film was chosen as $\theta = \frac{1}{2}$ and then

$s^{s,\ominus}_{oncfig} = 0$ so that, from (5.88) the standard differential molar entropy of adsorption is $-s^{g,\ominus}_{tr}$. In the non-localised model the standard state for the surface is taken as $\dot{a} = \dot{a}^{\ominus} = 4.08T \times 10^{-16}$ cm^2 molecule^{-1}.† The molar translational entropy $s^{s,\ominus}_{tr}$ of the monolayer corresponding to this coverage is calculated using a two-dimensional analogue of the Sackur–Tetrode equation. The differential entropy $\bar{s}^{s,\ominus}_{tr}$ may then be obtained using the relation $\bar{s}^{s,\ominus}_{tr} = s^{s,\ominus}_{tr} - R$ (see Everett, 1957).

The differential entropy of adsorption $\Delta_a \bar{s}$ which can be obtained from experimental adsorption isotherms by the use of (5.64) and (5.65), corresponds to a particular surface coverage θ and area per molecule $\dot{a}$, and must be converted to the standard values $\Delta_a \bar{s}^{\ominus}_n$ and $\Delta_a \bar{s}^{\ominus}_l$ for the non-localised and localised models respectively. Thus

$$\Delta_a \bar{s}^{\ominus}_n = \Delta_a \bar{s} + R \ln (\dot{a}^{\ominus}/\dot{a}) \tag{5.91}$$

and

$$\Delta_a \bar{s}^{\ominus}_l = \Delta_a \bar{s} + R \ln \frac{\theta}{1-\theta}. \tag{5.92}$$

The terms $R \ln \dot{a}^{\ominus}/\dot{a}$ and $R \ln [\theta/(1-\theta)]$ represent the differences between $\bar{s}^s$ at the coverage corresponding to $\dot{a}$ and θ, and the values in the two standard states (i.e. $\dot{a}^{\ominus}$ and $\theta = \frac{1}{2}$ respectively for the non-localised and localised models). The use of (5.91) and (5.92) clearly requires a knowledge of the surface area of the adsorbent. The standard differential entropies of adsorption given by (5.91) and (5.92) should be independent of coverage and can be compared with the values obtained theoretically for the two models. De Boer and Kruyer have examined many systems in this way and an example of some of the results is given in table 5.2. Also included in the table are values of $\Delta_a \overline{\mathscr{H}}$ which, according to both models, should be independent of θ. For argon adsorbed on charcoal at 215 K there is better agreement between values of $\Delta_a \bar{s}^{\ominus}_n$ and $(\bar{s}^{s,\ominus}_{tr} - s^{g,\ominus}_{tr})$ than between $\Delta_a \bar{s}^{\ominus}_l$ and $-s^{g,\ominus}_{tr}$. Thus in this case the adsorbed film behaves more nearly like an entropically ideal non-localised monolayer than a localised monolayer. However, the foregoing type of analysis does not necessarily give a true picture of the physical reality. As Everett (1957) points out, it is perhaps not surprising

† This choice of standard state corresponds to a temperature independent standard surface pressure, $\pi^{\ominus}$ ($= kT/\dot{a}^{\ominus}$) of 0.338 mN m^{-1}. It is chosen such that the average separation of the adsorbed molecules (in an assumed ideal two-dimensional film) is identical to the average separation in the gas at 0° C and 1 atmosphere. Such a choice is of course quite arbitrary but has been fairly widely used. However the particular choice of standard states does not influence the comparison of theoretical and experimental entropies in any way.

TABLE 5.2. *Differential heats and entropies* of adsorption of argon on charcoal at* 215 K

θ	$-\Delta_a\overline{\mathscr{H}}$	$-\Delta_a\bar{s}_l^{\ominus}$	$s_{tr}^{g,\ominus}$	$-\Delta_a\bar{s}_n^{\ominus}$	$s_{tr}^{g,\ominus}-\bar{s}_{tr}^{s,\ominus}$
0.16	14.46	69.9	147.9	36.8	55.5
0.20	14.63	70.3	—	38.5	—
0.24	14.50	70.3	—	38.5	—
0.26	14.46	70.7	—	38.5	—

The value in the last column differs by an amount *R* from that given by de Boer and Kruyer, who supposed that $\bar{s}_{tr}^{s,\ominus}$ and $s_{tr}^{s,\ominus}$ were equivalent.

* Entropy values are given in J mol^{-1} K^{-1}, and $\Delta_a\overline{\mathscr{H}}$ is in kJ mol^{-1}.

that a real film does not approximate to a model in which the vibrational entropy is assumed to be zero. If account is taken in the localised model of vibrations normal to and parallel to the surface, then closer agreement between experimental and theoretical entropies could be expected.

It is interesting to consider the variation of heats and entropies of adsorption with the surface coverage. Hill, Emmett & Joyner (1951) have discussed the general form of the entropy curves on the basis of the BET model (see §5.12). If the adsorbate is relatively strongly bound to the surface i.e. if c in (5.37) is relatively large, a minimum in s^s is predicted at $v = v_m$. The initial decrease in entropy at low coverages is mainly a result of the change in configurational entropy with surface concentration, and the increase in s^s for $v > v_m$ reflects the building up of multilayers. Some results for argon adsorbed on rutile at 85 K (where $c \simeq 100$), obtained by Drain & Morrison (1952) are shown in fig. 5.23. Both s^s and $\bar{s}^s$ approach the value for the entropy of liquid argon for high v ($\sim 3v_m$). When the adsorption was carried out below the normal melting point of argon (83.9 K) it was found that $\bar{s}^s$ approached the value for the entropy of solid argon at the appropriate temperature.

According to the BET model, in cases where the first layer is substantially complete before higher layers become populated, the differential heat (calorimetric or isosteric) should fall from an initially constant value, to the heat of liquefaction of the adsorbate at $v = v_m$, and thereafter remain constant. The heats of adsorption of nitrogen on graphitised carbon are shown in fig. 5.24. The fall in the heat at very low coverage may be due to surface heterogeneity, and the subsequent rise to lateral interactions within the film. Neither of these effects is allowed

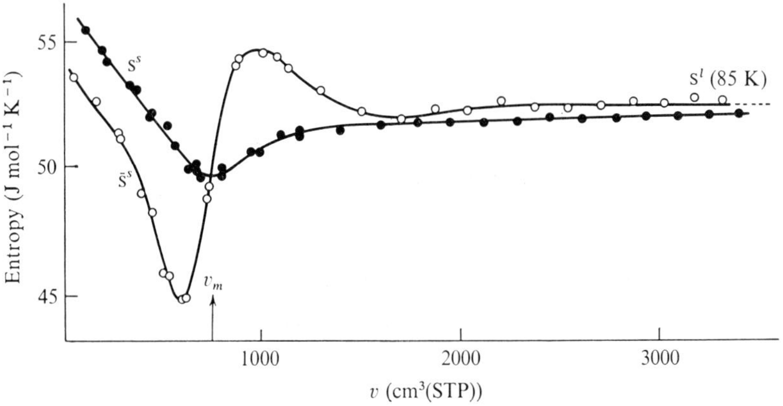

Fig. 5.23. The molar (s^s) and differential molar ($\bar{s}^s$) entropies of argon adsorbed on rutile at 85 K as a function of the volume adsorbed. (After Drain & Morrison, 1952.)

for in the BET theory. There is, however, as expected from BET theory, a sharp fall in the region of the completion of the monolayer, and the heat becomes constant and close to the heat of liquefaction of nitrogen.

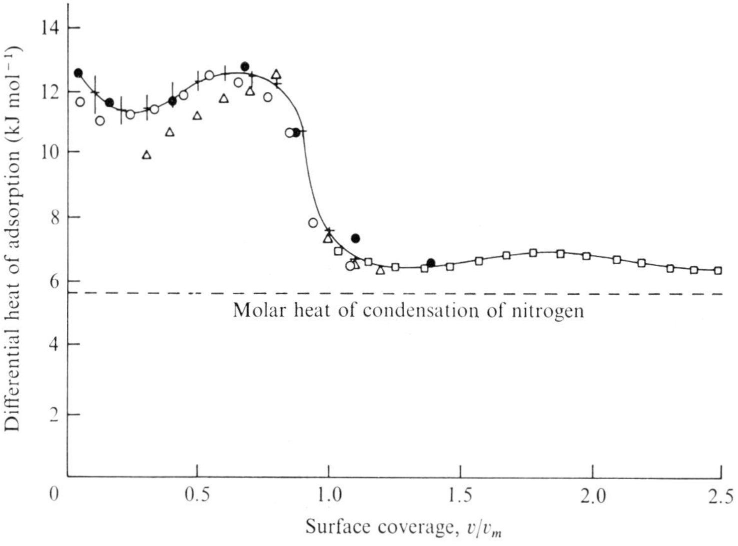

Fig. 5.24. Differential heats of adsorption of N_2 on Graphon. ○ and ● are calorimetrically determined; others are isosteric heats obtained from −194.8 to −182.8 °C (+), from −207.7 to −182.2 °C (△) and from −204.7 to −194.8 °C (□). (After Joyner & Emmett, 1948.)

Appendix

Thermodynamic functions of adsorbed layers and the concept of the inert solid in gas–solid adsorption

In §5.22 the extensive thermodynamic properties for the surface, e.g. A^s, are defined by equations of the type

$$A^s = A - A_0 \quad \text{etc.} \tag{5.93}$$

where A is the Helmholtz free energy of the system comprising n_A moles of adsorbent A and n^s moles of adsorbed gas. A_0 is the free energy of n_A moles of A in exactly the same state of subdivision and with the same specific surface area, but in the absence of the adsorbed gas. The above definition of surface properties may be used whether or not the solid is perturbed by the presence of adsorbed gas. In the special case where the thermodynamic properties of the solid are unaffected by the adsorbed gas however, i.e. for *inert* solids, the quantities A^s etc. can be associated entirely with the adsorbed gas. In this instance A^s is identical to A^{ads} used in §1.10. It will be recalled that $A^{\text{monolayer}}$ used in §3.7 is defined in a similar fashion to A^s. However, $A^{\text{monolayer}}$ will normally contain a contribution from the liquid substrate even though this is sometimes assumed to be negligible.

Consider the differential expression for the internal energy U of the system of adsorbent plus adsorbed gas,

$$\mathrm{d}U = T\mathrm{d}S - p\mathrm{d}V + \mu_A \mathrm{d}n_A + \mu \mathrm{d}n^s, \tag{5.94}$$

where μ_A and μ are, respectively, the chemical potentials of the adsorbent and the adsorbed gas. Changes, $\mathrm{d}n_A$ in n_A are brought about by adding A in exactly the same form as that originally present. It will be appreciated therefore that the interfacial area $\mathscr{a}$ of the system is proportional to n_A, and is thus not an independent variable. In the absence of adsorbed gas, (5.94) is written

$$\mathrm{d}U_0 = T\mathrm{d}S_0 - p\mathrm{d}V_0 + \mu_{A_0} \mathrm{d}n_A$$

so that
$$\mathrm{d}U^s = T\mathrm{d}S^s - p\mathrm{d}V^s + (\mu_A - \mu_{A_0})\mathrm{d}n_A + \mu \mathrm{d}n^s, \tag{5.95}$$

where U^s, S^s and V^s are defined in an analogous way to A^s in (5.93).

It is now convenient to express μ_A and μ_{A_0} each as the sum of two terms, i.e.

$$\mu_A = \mu'_A + \gamma a_A, \qquad \mu_{A_0} = \mu'_{A_0} + \gamma^0 a_A, \tag{5.96}$$

where μ'_A and μ'_{A_0} are the (hypothetical) chemical potentials of A assuming it had no surface area, and a_A ($= \Sigma M$) is the molar surface area of A. Equations (5.96) are equivalent to (1.33). If the solid is inert then $\mu'_A = \mu'_{A_0}$. The chemical potential of dispersed material was expressed in the form of (5.96) in the derivation of the Kelvin equation. The term $2V_\gamma/r$ in (3.7) is the partial molar area of component *i* which is added to an existing number of droplets of radius *r*. In the solid–gas system, A is added in the form of extra particles of the same specific surface area as the solid already present.

From (5.95) and (5.96) for an inert solid

$$dU^s = TdS^s - pdV^s + (\gamma - \gamma^0)a_A dn_A + \mu dn^s$$

or

$$dU^s = TdS^s - pdV^s - \pi d\mathcal{A} + \mu dn^s$$

which is (5.57).

References

Adamson, A. W. (1967). *Physical Chemistry of Surfaces*, 2nd edition (Interscience, New York): chapter 5.

Benson, G. C. & Yun, K. S. (1967). *The Solid–Gas Interface* (Arnold, London): vol. 1, chapter 8.

Brunauer, S. (1944). *The Adsorption of Gases and Vapours* (Oxford University Press): vol. 1, (*a*) pp. 104 *et seq.*, (*b*) p. 154; (*c*) p. 168.

Brunauer, S., Emmett, P. H. & Teller, E. (1938). *J. Amer. Chem. Soc.* **60**, 309.

Brunauer, S., Deming, L. S., Deming, W. E. & Teller, E. (1940). *J. Amer. Chem. Soc.* **62**, 1723.

de Boer, J. H. & Kruyer, S. (1952). *Koninkl. Ned. Akad. Wetenschap. Proc.* **55** B, 451,

de Boer, J. H. (1968). *The Dynamical Character of Adsorption*, 2nd edition (Clarendon Press): chapter 3.

Drain, L. E. & Morrison, J. A. (1952). *Trans. Faraday Soc.* **48**, 840.

Drake, L. C. & Ritter, H. L. (1945). *Ind. Eng. Chem.* (*Anal*). **17**(12), 787.

Dubinin, M. M. (1955). *Quart. Rev.* **9**, 101.

Dubinin, M. M. (1960). *Chem. Rev.* **60**, 235.

Emmett, P. H. (1942). *Adv. in Colloid Sci.* (Interscience, New York): vol. 1, pp. 1–36.

Everett, D. H. (1950). *Trans. Faraday Soc.* **46**, 453.

Everett, D. H. (1957). *Chem. Soc. Proc.* 38.

Flood, E. A. (editor) (1967). *The Solid–Gas Interface* (Arnold): vol. 1.

Gregg, S. J. (1965). *The Surface Chemistry of Solids* (Chapman and Hall): (*a*) pp. 137–42; (*b*) p. 286.

Guggenheim, E. A. (1952). *Mixtures* (Clarendon Press): chapter 4.

Halsey, G. D. (1948). *J. Chem. Phys.* **16**, 931.

Hill, T. L. (1946). *J. Chem. Phys.* **14**: (*a*) 441; (*b*) 263.

Hill, T. L. (1952). *Adv. Catalysis*, **4**, 211 (Academic Press, New York).

Hill, T. L., Emmett, P. H. & Joyner, L. G. (1951). *J. Amer. Chem. Soc.* **73**, 5102.

Johnson, R. E. (1959). *J. Phys. Chem.* **63**, 1655.

Joyner, L. G. & Emmett, P. H. (1948). *J. Amer. Chem. Soc.* **70**, 2353.

Kälberer, W. & Schuster, C. (1929). *Z. phys. Chem.* A **141**, 270.

Kemball, C. (1950). *Adv. Catalysis*, **2**, 233 (Academic Press, New York).

London, F. (1930). *Z. phys. Chem.* **11**, 222.
McBain, J. W. (1935). *J. Amer. Chem. Soc.* **57**, 699.
Polanyi, M. (1914). *Verh. ditsch. phys. Ges.* **16**, 1012.
Ritter, H. L. & Drake, L. C. (1945). *Ind. Eng. Chem.* (*Anal.*), **17**(12), 782.
Ross, S. & Clark, H. (1954). *J. Amer. Chem. Soc.* **76**, 4291.
Ross, S. & Olivier, J. P. (1964). *On Physical Adsorption* (Interscience): experimental techniques are discussed in chapter 2.
Steele, W. A. (1967). *Adv. Colloid and Interface Sci.* **1**, 3.
Taylor, H. S. (editor) (1931). *Treatise on Physical Chemistry*, 2nd edition (McMillan London): vol. 2; chapter 20 by E. O. Kraemer, p. 1661.
Titoff, A. (1910). *Z. phys. Chem.* **74**, 641.
Wayman, R. A. (1967). M.Sc. thesis, Hull University.
Young, D. M. & Crowell, A. D. (1962). *Physical Adsorption of Gases* (Butterworths): (*a*) chapter 2; (*b*) pp. 159–63; (*c*) chapter 5; (*d*) chapter 3.
Zsigmondy, A. (1911). *Z. anorg. Chem.* **71**, 356.

6 The solid–liquid interface

6.1. Introduction. The following discussion of solid–liquid interfaces is concerned first with the wetting of solids by pure liquids but thereafter, and mainly, with adsorption from liquid mixtures and dilute solutions.

The study of adsorption from solution on to solids is very wide-ranging and many of the practical uses have long been known. The solutions can be broadly subdivided into dilute solutions and liquid mixtures. For dilute solutions the solutes may be ionic (important examples being certain classes of dyes and detergents) or non-ionic (e.g. polymers and aliphatic polar compounds). Much of the work on liquid mixtures has involved the use of binary systems of miscible organic liquids. In recent years there has been a renewed interest in adsorption from liquid mixtures. It is particularly interesting to point to the similarities in the theory of adsorption from mixtures at the liquid–vapour and the solid–liquid interfaces. Although such similarities are to be expected they have, in the past, been somewhat obscure.

6.2. Wetting and adhesion in solid–liquid systems: basic concepts. In the following discussion effects due to gravity and curvature have been excluded and use will be made of the definitions and concepts discussed in §3.5.

Fig. 6.1 depicts a drop of liquid α at equilibrium on the plane surface of solid β. The contact angle θ is a measurable quantity in solid–liquid systems. The superscripts α and β refer to the liquid–vapour and the solid–vapour interfaces respectively. Young's equation is

$$\gamma^{\beta} = \gamma^{\alpha\beta} + \gamma^{\alpha} \cos\theta, \tag{3.22}$$

where the γs are defined by

$$\gamma = (\partial A/\partial \mathscr{A})_{T,V,n_i}. \tag{6.1}$$

The quantity A is the Helmholtz free energy of the system and $\mathscr{A}$ is the area of the appropriate interface. The notation γ is discussed in §5.4.

Combination of the Dupré equation (3.20) and Young's equation

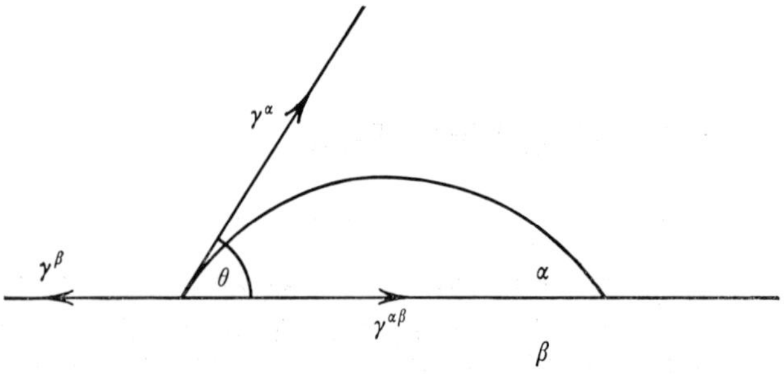

Fig. 6.1. A drop of liquid α resting on the surface of solid β.

(3.22) leads to the relationship for the work of adhesion W_A between the solid and the liquid,

$$W_A = \gamma^{\alpha}(1+\cos\theta). \tag{6.2}$$

It is therefore possible to calculate W_A from the measurable quantities γ^{α} and θ. The experimentally inaccessible quantity $\gamma^{\alpha\beta}$ is not required. Equation (6.2) is applicable to a system at equilibrium and so it is necessary that the components of the system should have uniform chemical potential in the liquid, solid and vapour phases. For this reason, the liquid must be saturated with the solid, which in general leads to a lowering of the surface tension of the liquid, and the vapour and solid surface must be at adsorption equilibrium. This latter requirement often means that the solid surface is covered by a multimolecular film of α. The value of γ^{β} for the film-covered surface is lower than that ($\gamma^{0,\beta}$) for the solid–vacuum interface by an amount π, the surface pressure of the adsorbed film, i.e.

$$\gamma^{\beta} = \gamma^{0,\beta} - \pi. \tag{6.3}$$

It follows that W_A in (6.2) is the work required to part unit area of solid and liquid, the final solid surface having on it the equilibrium adsorbed film. Combination of (3.22) and (6.3) shows that Young's equation may be expressed

$$\gamma^{0,\beta} - \pi = \gamma^{\alpha\beta} + \gamma^{\alpha}\cos\theta. \tag{6.4}$$

The work of adhesion W_A^0 of a liquid and a clean solid surface is given by

$$W_A^0 = W_A + \pi. \tag{6.5}$$

π can be obtained experimentally from the appropriate adsorption data by use of the expression

$$\pi = RT\int_{p=0}^{p=p^0} \Gamma^s \,\mathrm{d}\ln p \tag{6.6}$$

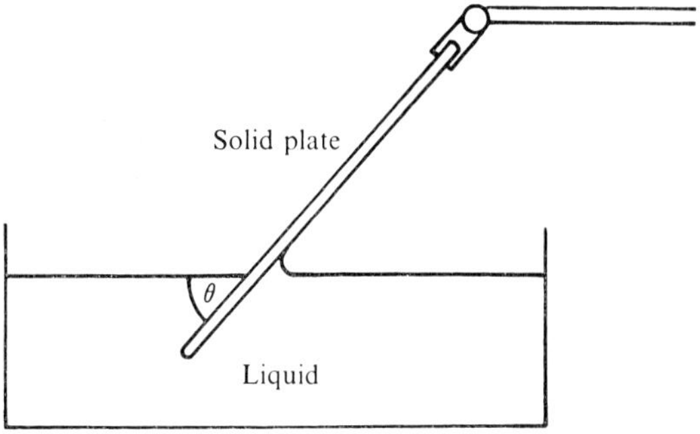

Fig. 6.2. The tilting plate method for the solid–liquid contact angle.

as discussed in §5.4. p^0 is the saturated vapour pressure of α at temperature T.

Initial and final spreading tensions, $\sigma_i^{\alpha\beta}$ and $\sigma^{\alpha\beta}$ respectively, can be defined and have the same significance as for liquid–liquid systems.

6.3. The contact angle θ. Several methods are available for the determination of θ and the technique chosen depends largely on the nature and availability of the solid and liquid to be studied. If it is possible to prepare a flat plate of the solid a few centrimetres in length, as for example with glass and certain metals, and if a sufficient volume of liquid is available, the *tilting plate* method can be employed. The plate is held in a clamp, which can be appropriately adjusted, and dipped into the liquid (fig. 6.2). The angle of the plate is arranged so that there is no curvature on the surface of the liquid in contact with one side of the plate, as depicted. The angle of the plate from the horizontal which is equal to θ is then determined.

Another method, the sessile drop technique, involves taking a photograph of a drop of the liquid resting on the solid surface. θ can be obtained either by direct inspection of the photograph or by the measurement of the dimensions of the drop. If the solid is in a powdered form (e.g. ores) indirect methods for the measurement of θ must be used.

Values of θ for a given system often vary widely. The discrepancies may be explained in a variety of ways. In some instances it is possible that equilibrium has not been attained. In other instances it is likely that contaminants were present. The presence of surface active impurities, such as grease or detergents, can drastically alter the value of θ. That this is so, is easily appreciated by noting that θ for pure water on clean glass is zero, whereas under normal conditions (e.g. a window pane when the glass has a layer of adsorbed hydrophobic impurity) water will not spread over the surface.

Hysteresis of the contact angle is commonly encountered. Suppose that the bath of liquid depicted in fig. 6.2 can be raised and lowered. If, initially, the bath is raised, liquid will come into contact with solid which has not previously been covered. θ determined under these circumstances is the *advancing contact angle*. If now the bath is lowered, the line of solid–liquid contact occurs at a position on the solid which has previously been immersed. θ obtained in this way is called the *receding contact angle*. Hysteresis is said to occur when values of the advancing and receding contact angles for a given system are not equal. Several factors, including surface roughness and the presence of surface impurities can cause hysteresis. For a more detailed discussion of phenomena involving contact angles, including hysteresis and surface roughness, the reader is recommended to consult Gould (1964).

6.4. Wetting and surface constitution. The values of the contact angle, work of adhesion and spreading tension (see §3.5) for a system will clearly be a function of the nature of the solid surface. An interesting example is furnished by the work of Adam & Elliott (1962). These authors determined the contact angles of water with several solid saturated hydrocarbons, some of which expose only methyl groups at the surface (e.g. hexamethylethane, $\theta = 115 \pm 3°$) and some where only methylene groups are exposed (e.g. cycloparaffins with 15, 16 and 17 $-CH_2-$ groups, $\theta = 104.5 \pm 1°$). Substitution of these values of the contact angles in (6.2) gives the work of adhesion of water with the $-CH_3$ group as between 38 and 42 mJ m^{-2} and with the $-CH_2-$ group as between 53 and 56 mJ m^{-2}. Minor differences in surface constitution can therefore lead to marked differences in the values of interfacial properties.

By means of the so-called *retraction method* solid surfaces can be prepared which expose desired chemical groupings (Zisman, 1964). A slide of, say, polished platinum foil is dipped into either a dilute solution of a polar paraffinic substance in an organic solvent, or into the pure molten substance. The plate is then retracted from the liquid when it is found that the surface is covered by a condensed monolayer of the paraffinic material. The results of a study of wettability using surfaces formed in this way, comprising close-packed $-CF_3$, $-CF_2H$, $-CF_2-$ and $-CH_3$ groups are shown in fig. 6.3. Zisman and co-workers found empirically that, for a given solid surface, there often exists a

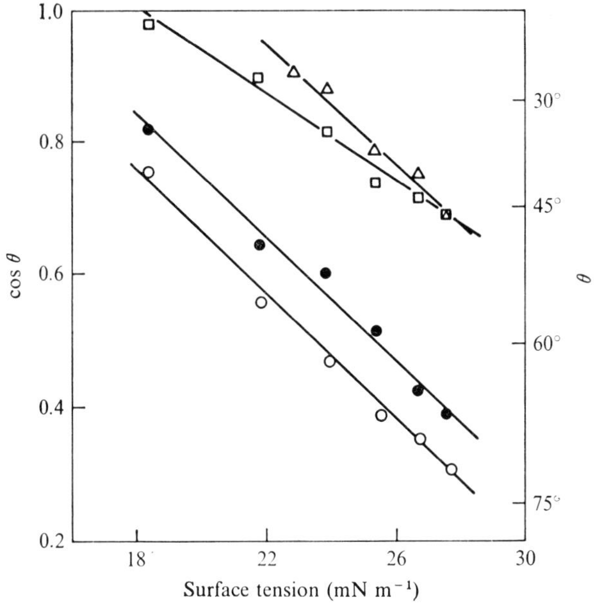

Fig. 6.3. Variation of contact angle with surface tension for n-alkanes on surfaces exposing $-CF_3$ (○), $-CF_2H$ (●), $-CF_2-$ (□), and $-CH_3$ (△). (From Ellison, Fox & Zisman, 1953.)

rectilinear relationship between cos θ and the surface tension of the members of an homologous series of liquids. The liquids used in the present example were n-alkanes. It can be seen that the wettability of the surface groups is $-CH_3 > -CF_2- > -CF_2H > -CF_3$.

A prior knowledge of contact angles between liquids and solid surfaces of known constitution has been of considerable use in the preparation of textile fibres and fabrics with specific oil- or water-repellant properties.

6.5. Adsorption from binary liquid mixtures of non-electrolytes: introduction. When an insoluble solid is immersed in a solution there is usually a change in composition of the solution. This is a result of the preferential adsorption of one of the components. The practical use of adsorption from solution on to solids has long been appreciated, but the thermodynamic aspects have received far less attention than has the thermodynamics of the adsorption of gases on solids.

Recently Schay and co-workers (see Schay *et al.* 1962) and Everett

(1964, 1965) have drawn attention to the similarities between the adsorption from binary liquid mixtures at the liquid–vapour and the liquid–solid interfaces. In both cases the surface coverage is always complete and only the composition of the surface changes when the bulk composition is changed. This is in contrast to the adsorption of gases on solids, where the surface coverage is a function of the gas pressure. There are of course differences in the experimental techniques used for the study of adsorption at liquid–vapour and liquid–solid interfaces. For the former the adsorption is determined from surface tension data (§1.8) whereas for the latter this is not possible and the adsorption is obtained more directly from the measurement of changes in the composition of the liquid.

When an adsorbent is added to a binary solution it is clear that both components will be present at the interface, and the change in compostion of the liquid is therefore a result of the adsorption (positive or negative) of both the components. The isotherm which represents this adsorption is termed a *surface excess isotherm* or a *composite isotherm*. *Individual isotherms* for the adsorption of each species separately can be obtained from the surface excess isotherm if further information is available (see §6.7).

6.6. The surface excess isotherm for adsorption from binary liquid mixtures. The following notation will be used:

n_0	the total number of moles of solution before adsorption.
$n_{1,0}, n_{2,0}$	the total number of moles of components 1 and 2 in solution before adsorption.
n_1, n_2	the number of moles of components 1 and 2 in solution at adsorption equilibrium.
$x_{1,0}, x_{2,0}$	the mole fractions of components 1 and 2 in solution before adsorption.
x_1, x_2	the mole fractions of components 1 and 2 in solution at adsorption equilibrium.
n_1^s, n_2^s	the numbers of moles of components 1 and 2 on the surface of unit mass of solid at adsorption equilibrium.†

The system envisaged consists of a mass m_A of insoluble solid adsorbent which is added to n_0 $(= n_{1,0}+n_{2,0})$ moles of solution. The system

† Elsewhere in this book n_i^s is used to denote the number of moles in any area $\mathcal{A}$. In this chapter n_i^s is the number of moles in the specific surface area Σ, i.e. on unit mass of solid. Since the symbol is used *exclusively* in this sense in chapter 6 it is felt unnecessary to introduce a new symbol.

is allowed to attain adsorption equilibrium. The amounts of components 1 and 2 present in the system are the same before and after adsorption has occurred. It follows therefore that

$$m_A n_1^s + n_1 = n_{1,0} \tag{6.7}$$

and

$$m_A n_2^s + n_2 = n_{2,0}. \tag{6.8}$$

Since $n_1/n_2 = x_1/x_2$, (6.7) and (6.8) may be expressed respectively as

$$m_A n_1^s + n_2 x_1/x_2 = n_{1,0} \tag{6.9}$$

and

$$m_A n_2^s + n_1 x_2/x_1 = n_{2,0}. \tag{6.10}$$

Multiplication of (6.9) by x_2 and (6.10) by x_1 gives

$$m_A n_1^s x_2 + n_2 x_1 = n_{1,0} x_2 \tag{6.11}$$

and

$$m_A n_2^s x_1 + n_1 x_2 = n_{2,0} x_1. \tag{6.12}$$

Subtraction of (6.12) from (6.11), noting that $n_2 x_1 = n_1 x_2$ yields

$$m_A(n_1^s x_2 - n_2^s x_1) = n_{1,0} x_2 - n_{2,0} x_1. \tag{6.13}$$

Substitution for x_2 and $n_{2,0}$ on the right-hand side of (6.13), using $x_2 = 1 - x_1$, and $n_{2,0} = n_0 - n_{1,0}$ and noting at the same time that $n_{1,0} = n_0 x_{1,0}$, leads to the equation for the surface excess isotherm

$$n_0 \Delta x_1/m_A = n_1^s x_2 - n_2^s x_1, \qquad \blacktriangleleft\mathbf{(6.14)}$$

where $\Delta x_1 = x_{1,0} - x_1$. $n_0 \Delta x_1/m_A$ can be obtained experimentally. Δx_1 is commonly determined by observing the change in refractive index of the solution, produced by adsorption. The data are normally presented as plots of $n_0 \Delta x_1/m_A$ against x_1. Positive values of $n_0 \Delta x_1/m_A$ (i.e. $x_{1,0} > x_1$) indicate positive adsorption of component 1. It then follows that $n_0 \Delta x_2/m_A$ is negative, i.e. component 2 is negatively adsorbed and the surface phase is less rich than the bulk in component 2. It is obvious that the adsorption must be zero for the two pure components. The isotherm (6.14) is clearly a composite isotherm since both n_1^s *and* n_2^s appear on the right-hand side of the equation.

For completely miscible liquid pairs neither of the components is termed the solvent or solute, and experimental isotherms normally extend over the whole range of composition from $x_1 = 0$ to $x_1 = 1$.

In general, surface excess isotherms are either U-shaped or S-shaped, as shown in fig. 6.4. U-shaped isotherms result from the preferential adsorption of the same component at all bulk compositions, whereas

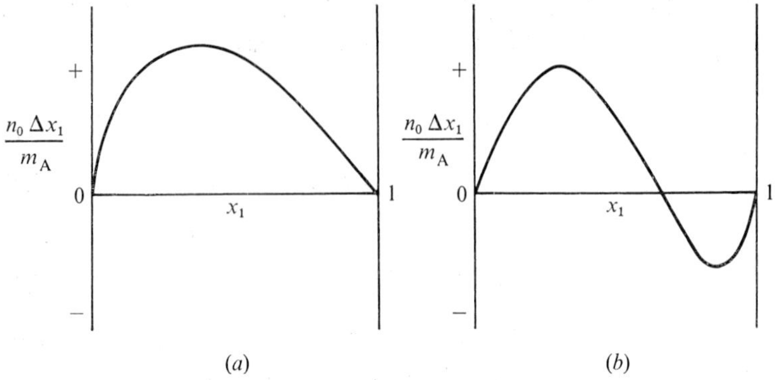

Fig. 6.4. (*a*) U-shaped isotherm. (*b*) S-shaped isotherm.

the S-shaped isotherms are obtained when a given component is positively adsorbed over part of the composition range, and negatively adsorbed over the remainder. An example of a U-shaped isotherm is found for boehmite ($\gamma-Al_2O_3.H_2O$) in mixtures of chloroform and benzene where the chloroform is positively adsorbed. Adsorption from ethanol+benzene mixtures on to Spheron 6 (a carbon black with a heterogeneous surface) gives an S-shaped isotherm. Ethanol is positively adsorbed for mole fractions of ethanol from 0 to approximately 0.4 whereas the benzene is preferentially adsorbed from mole fractions of 0.4 to 1.

The shape of a surface excess isotherm may be influenced by the nature of the adsorbent surface and by the bulk solution properties. For a system in which the binary mixtures were perfect and the adsorbent surface homogeneous, a U-shaped isotherm would be expected. S-shaped isotherms are frequently observed for adsorption on various forms of carbon (Kipling, 1965*a*). The surfaces of the latter often contain oxygen complexes, which endow heterogeneity. Thus, for example, it may be that in mixtures of polar organic molecules with benzene, the polar molecule interacts strongly with the surface oxygen complexes giving rise to preferential adsorption of the polar material at low concentrations. When the polar sites are saturated, the benzene is preferentially adsorbed on the remaining surface. An example of this effect is seen in the adsorption from mixtures of n-butanol with benzene on Spheron 6, which gives an S-shaped isotherm, the butanol being preferentially adsorbed at low butanol concentrations, and the benzene preferentially adsorbed at higher concentrations. On Graphon, however, which is a relatively homogeneous carbon (§5.2) benzene is preferentially adsorbed at all concentrations. It can also be shown that adsorption from regular solutions on homogeneous surfaces can result in S-shaped isotherms, but a discussion of this problem is beyond the scope of this presentation.

It has been pointed out that $n_0\Delta x_1/m_A$ reflects the adsorption of both components in a mixture. However, there are special cases in which the adsorption of one species alone may be obtained. An example of this is the adsorption from dilute solutions, which is treated in §6.9. Another interesting case arises for liquid mixtures in contact with molecular sieves. Zeolites, which consist of networks of $[Si, Al]_nO_{2n}$ with balancing cations (e.g. Na^+ or K^+), are examples of molecular sieves, and have a three-dimensional porous structure. Synthetic 'A-type' sieves include cavities of about 1.2 nm diameter, which are connected by 'windows' with diameters of between 0.3 and 0.5 nm, depending on the nature of the cation (Minkoff & Duffett, 1964). When such sieves are brought in contact with a mixture of molecules, only those species which are sufficiently small to diffuse through the 'windows' and into the cavities are taken up by the sieve, which acts like a sponge. Molecular sieves are used commercially to separate normal paraffins from other hydrocarbons which have larger molecular cross-sectional areas.

Now suppose that a molecular sieve is added to a binary liquid mixture and that one of the components (say 1) is 'adsorbed' exclusively. Then $n_2^s = 0$ and (6.14) becomes

$$n_0\Delta x_1/m_A = n_1^s x_2. \tag{6.15}$$

If it is supposed that the cavities are filled at all liquid compositions (except of course for $x_2 = 1$), then n_1^s is constant and $n_0\Delta x_1/m_A$ is a linear function of x_2. Such behaviour has been observed for example for the system benzene+n-hexane in contact with Linde molecular sieve 5A. n-Hexane, which has an effective molecular diameter of 0.49 nm, is sorbed whereas benzene (molecular diameter 0.63 nm) is excluded.

6.7. The individual isotherm. In general, values of $n_0\Delta x_1/m_A$ are not sufficient for the calculation of n_1^s and n_2^s separately. Kipling (1965*b*) has discussed various methods which may be used to obtain the necessary additional information. When the adsorption from the liquid mixture is identical to that which takes place from the mixed vapour in equilibrium with the liquid, the information can be obtained experimentally. The adsorption from the vapour is determined gravimetrically. Suppose W is the increase in mass of unit mass of adsorbent, resulting from adsorption, then

$$W = n_1^s M_1 + n_2^s M_2, \tag{6.16}$$

where M_1 and M_2 are the molecular weights of components 1 and 2 respectively. Combination of (6.14) and (6.16) allows the calculation of n_1^s and of n_2^s to be made. However, it may be that adsorption from the liquid and vapour phases is not the same. For example, adsorption from the vapour on to a porous solid may involve capillary condensation (§5.15). The condensate will clearly contribute to the value of W, but is not present as a result of the primary adsorption process.

Other approaches to obtaining individual isotherms are also possible. One of the simplest and most widely used is to make the assumption that the adsorbed layer is monomolecular. This same assumption is frequently made when treating adsorption from liquid mixtures at the liquid–vapour interface (see §1.8). The assumption leads to the equation

$$\frac{n_1^s}{(n_1^s)_m} + \frac{n_2^s}{(n_2^s)_m} = 1, \tag{6.17}$$

where $(n_1^s)_m$ and $(n_2^s)_m$ are, respectively, the numbers of moles of components 1 and 2 required, individually, to complete a monolayer on unit mass of solid. Equation (6.17) is equivalent to (1.47) which was applied in the form of (3.28) to the liquid–vapour interface. The quantities $(n_1^s)_m$ and $(n_2^s)_m$ are directly related to the respective monolayer capacities discussed in §5.9 and, if it is assumed that the orientations of the molecules are the same at the solid–vapour and the solid–liquid interfaces, they may be obtained from separate vapour adsorption experiments (§5.9 and §5.12). Adsorption of the individual components from the liquid mixture can be calculated using (6.17) and (6.14).

It is not possible to prove that an adsorbed layer at a solid–liquid interface is monomolecular for a given system. It can be seen from (6.14), however, that a necessary condition for monolayer adsorption is that for all values of x_1

$$n_0 \Delta x_1 / m_A \ngtr (n_1^s)_m x_2 \tag{6.18}$$

since the highest possible value of $n_0 \Delta x_1 / m_A$ occurs for $n_1^s = (n_1^s)_m$ and $n_2^s = 0$. If, for any system (6.18) does not hold, then (6.17) cannot be used.

A procedure for the estimation of the thickness of an adsorbed layer, which may be applied in cases where there is a linear section in the surface excess isotherm, has been suggested by Schay & Nagy (1961). In

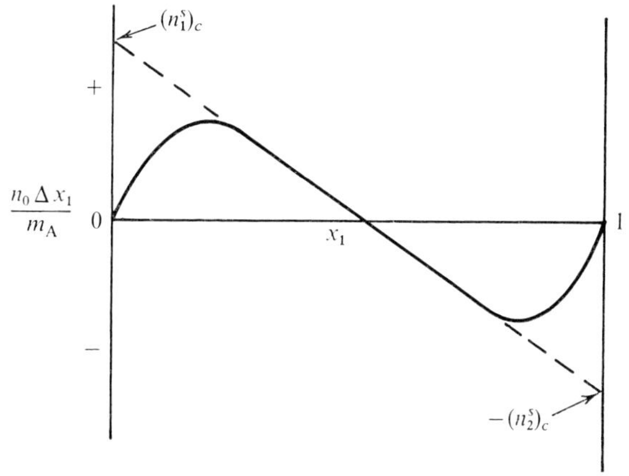

Fig. 6.5. Composite isotherm with a linear section.

the linear region, depicted in fig. 6.5, (6.14) represents a straight line. The linearity is consistent with the composition of the surface remaining unchanged over this range of bulk composition. This can be seen by noting that (6.14) may be written in the form

$$\frac{n_0 \Delta x_1}{m_A} = n_1^s - (n_1^s + n_2^s)\, x_1 \tag{6.19}$$

which, if n_1^s and n_2^s are constant, is the equation of a straight line. The constant values of n_1^s and n_2^s, denoted $(n_1^s)_c$ and $(n_2^s)_c$ are obtained from the intercepts at $x_1 = 0$ and $x_1 = 1$ as shown in fig. 6.5. Monolayer adsorption is indicated if

$$\frac{(n_1^s)_c}{(n_1^s)_m} + \frac{(n_2^s)_c}{(n_2^s)_m} = 1. \tag{6.20}$$

One way in which a surface phase of constant composition could occur is when the adsorbent has a heterogeneous surface with two types of adsorption site (*A* and *B*). Sites *A* may be readily saturated by component 1, whilst sites of type *B* may preferentially attract component 2. It is possible in such an instance to obtain, in the middle of the bulk composition range, a surface of constant composition, sites *A* and *B* being saturated with components 1 and 2 respectively The results for the adsorption from mixtures of various aliphatic alcohols and benzene on a steam activated coconut shell charcoal can be explained in this way. The adsorbent contains surface oxygen complexes which may preferentially adsorb alcohol molecules. The surface excess adsorption isotherms for methanol and for ethanol with benzene are shown in fig. 6.6(*a*); the approximately linear section extends over a fairly wide composition range. The individual isotherms for the system methanol and benzene are shown in fig. 6.6(*b*). It is seen that the adsorption of the components is constant in the middle of the range of bulk composition.

So far the discussion has centred on the possibility of monolayer adsorption. Multilayer formation, which can be detected as described earlier in the section is, however, common in adsorption from solutions of solids close to saturation, and from incompletely miscible liquid pairs close to the miscibility limit. It is probable, for example, that benzoic acid adsorbed from aqueous solution to on Graphon is present at the surface as a multilayer at bulk concentrations of about 2.4×10^{-2} M. Methanol close to its solubility limit in n-heptane appears to give multilayer adsorption on silica gel.

6.8. Some comparisons with adsorption from liquid mixtures at the liquid–vapour interface. Possibly one of the most useful advances in the theoretical treatment of adsorption at the solid–liquid interface in recent years has been the clarification of the similarities which exist with the liquid–vapour interface. The following discussion is intended to illustrate some of these similarities and to show how experimental results obtained for the two types of interface may be compared.

The surface excess isotherm (6.14) refers to adsorption on unit mass of solid. Division throughout by Σ, the specific surface area of the solid, gives

$$\frac{n_0 \Delta x_1}{m_A \Sigma} = \Gamma_1^s x_2 - \Gamma_2^s x_1, \tag{6.21}$$

where Γ_1^s $(= n_1^s/\Sigma)$ and Γ_2^s $(= n_2^s/\Sigma)$ are surface concentrations. A form of the surface excess isotherm applicable to adsorption at the liquid–vapour interface is (see 1.44)

$$-\frac{d\gamma}{d\mu_1} = \Gamma_1^s - \frac{x_1}{x_2}\Gamma_2^s = \Gamma_1^{(2)}. \tag{6.22}$$

Therefore

$$-x_2 \frac{d\gamma}{d\mu_1} = x_2 \Gamma_1^{(2)} = x_2 \Gamma_1^s - x_1 \Gamma_2^s. \tag{6.23}$$

The term $x_2 \Gamma_1^{(2)}$, like $\Gamma_1^{(2)}$, is a surface excess (see §1.8) and is written $\Gamma_1^{(N)}$. It is the number of moles of component 1 in unit area of surface less the number of moles of component 1 in that amount of bulk solu-

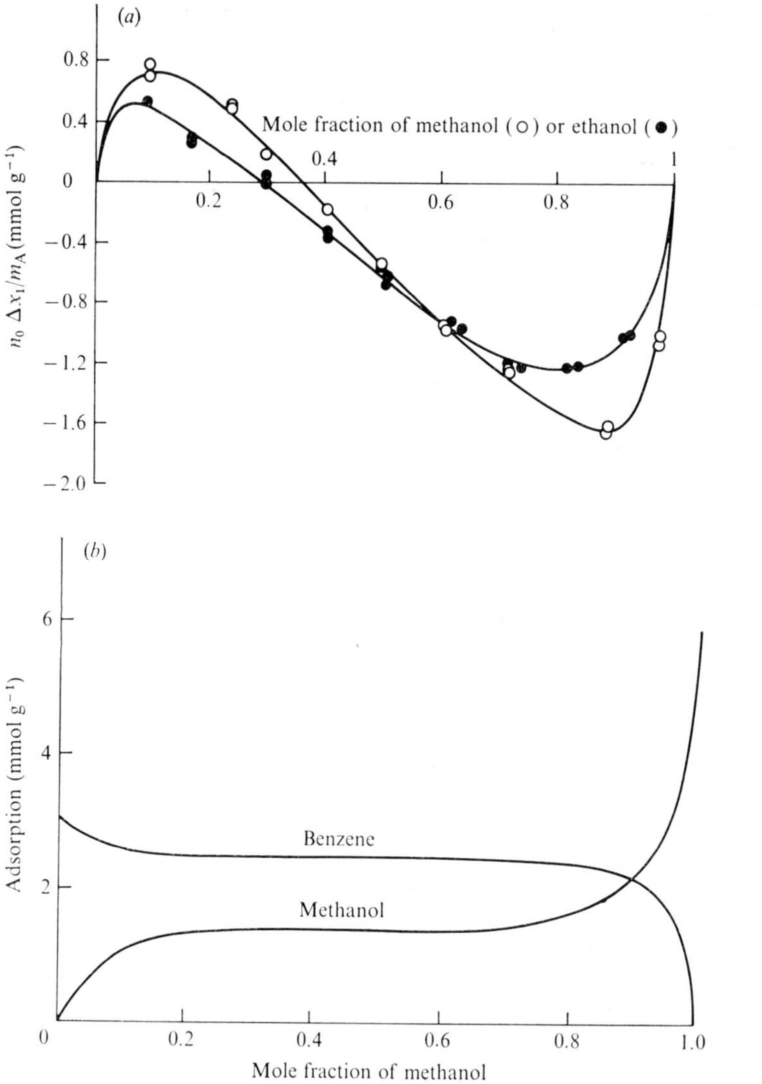

Fig. 6.6. (*a*) Composite isotherms for the adsorption from benzene and methanol (○) and benzene and ethanol (●) mixtures on an activated charcoal. (*b*) Individual adsorption isotherms for benzene and methanol adsorbed from benzene + methanol mixtures on an activated charcoal. (After Gasser & Kipling, 1960.)

tion which contains the same total number of moles (i.e. $\Gamma_1^s + \Gamma_2^s$) as unit area of surface. This can be appreciated by writing (6.23) as

$$-x_2 \frac{d\gamma}{d\mu_1} = \Gamma_1^{(N)} = \Gamma_1^s - x_1(\Gamma_1^s + \Gamma_2^s). \tag{6.24}$$

It is clear from the comparison of (6.21) and (6.23) that $n_0 \Delta x_1 / m_A \Sigma$ is a surface excess and has the same significance as $\Gamma_1^{(N)}$. For this reason it is very convenient to compare adsorption at the two types of interface in terms of these two quantities.

Langman (1965) has studied the adsorption from several binary liquid mixtures both at the liquid–vapour and at the liquid–Graphon interfaces. It will be recalled (§5.2) that Graphon is a relatively homogeneous non-porous form of carbon. The mixtures consisted of chemically similar liquids, and all the surface excess isotherms were found to be U-shaped. It can be seen from (6.23) that for the liquid–vapour interface it would be necessary for there to be a maximum or a minimum in the γ against activity isotherm for an S-shaped isotherm to result. It also follows that if a U-shaped isotherm is given, the component with the lower surface tension is preferentially adsorbed at the liquid–vapour interface. In all the systems studied by Langman it was the component with the higher surface tension which was preferentially adsorbed at the solid–liquid interface. This indicates that the liquids which have the highest surface tensions have the lowest values of γ for the Graphon–liquid interface. An example of the reversal of the preferential adsorption at the two types of interface is illustrated in fig. 6.7. In some systems, however, the preferentially adsorbed component is the same at the two types of interface. This is so, for example, for the adsorption from mixtures of benzene and chloroform at the liquid–vapour and the boehmite–liquid interfaces. Chloroform is preferentially adsorbed at both interfaces, but much more strongly so at the solid–liquid interface.

An alternative procedure for the comparison of adsorption data is to inspect the γ against composition data for the two interfaces. It has already been pointed out that values of γ for the solid–liquid interface are not directly accessible. The change, $\gamma_1^0 - \gamma$, which results when a mixture in contact with the solid is replaced by pure component 1 can, however, be computed from adsorption data. For the solid–liquid interface, (6.24) may be written

$$\frac{n_0 \Delta x_1}{m_A \Sigma} = -\frac{x_2}{RT} \frac{d\gamma}{d \ln \boldsymbol{a}_1}, \tag{6.25}$$

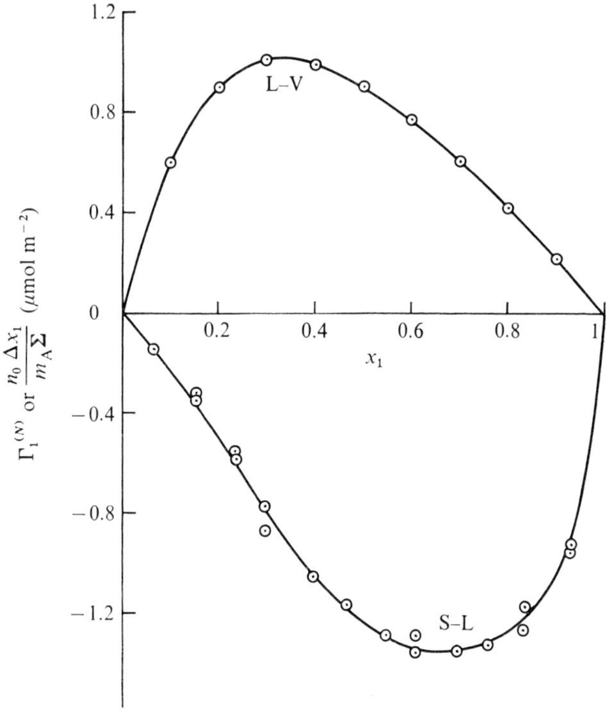

Fig. 6.7. Adsorption from cyclohexane (1)+chlorobenzene (2) mixtures at 20 °C, at the liquid–vapour (L–V) and Graphon–liquid (S–L) interface. (Results from Langman, 1965.)

where $\boldsymbol{a}_1$ is the mole fraction activity of component 1 in the mixture, and is replaced by x_1 for ideal mixtures. Rearrangement of (6.25) yields

$$-\mathrm{d}\gamma = \frac{n_0 \Delta x_1}{m_A \Sigma} \frac{RT}{x_2} \mathrm{d} \ln \boldsymbol{a}_1. \tag{6.26}$$

Integration of (6.26) then gives

$$\gamma_1^0 - \gamma = \frac{RT}{\Sigma} \int_{\boldsymbol{a}_1=1}^{\boldsymbol{a}_1} \frac{n_0 \Delta x_1}{x_2 m_A} \mathrm{d} \ln \boldsymbol{a}_1 \tag{6.27}$$

so that $\gamma_1^0 - \gamma$ may be obtained by the graphical integration of a plot of $n_0 \Delta x_1 / x_2 m_A$ against $\ln \boldsymbol{a}_1$. Results taken from the work of Nagy (1963) for the adsorption from mixtures of 1,2-dichloroethane (1) and benzene (2) are shown in fig. 6.8. The reversal of preferential adsorption is once again observed. 1,2-Dichloroethane is positively adsorbed on the

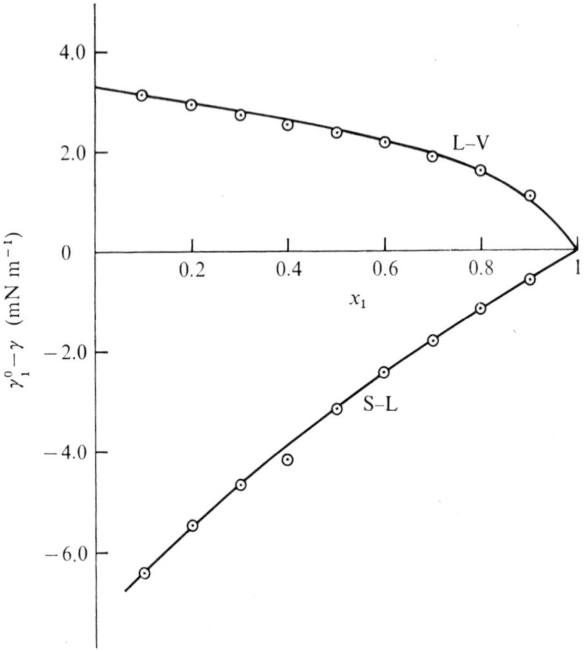

Fig. 6.8. $(\gamma_1^0-\gamma)$ for adsorption from mixtures of 1,2-dichloroethane (1) and benzene (2) at 25 °C, at the liquid–vapour (L–V) and liquid–silica gel (S–L) interface. (Results from Nagy, 1963.)

silica gel whereas benzene is preferentially adsorbed at the liquid–vapour surface. Schay *et al.* (1962) have pointed out that equations of the type of (3.35), originally developed for monomolecular adsorption at the liquid–vapour interface, should in principle be valid for monomolecular adsorption at the solid–liquid interface. For a perfect binary mixture in which the two components have the same molar surface areas, a, (3.35) may be written

$$\frac{x_1^m x_2^l}{x_1^l x_2^m} = \exp\left[\frac{(\gamma_2^0-\gamma_1^0)\,a}{RT}\right] = K. \tag{6.28}$$

Superscript l has been written above bulk mole fractions, as in chapter 3, to distinguish these quantities from the surface mole fractions x^m. x^l is equivalent to x used elsewhere in this chapter. Equation (6.28) should be applicable to the solid–liquid interface provided that the solid surface is homogeneous. The quantities γ_1^0 and γ_2^0 are not experimentally determinable for solids, but the constancy of K can be tested using values of x_1^m and x_2^m calculated as described in §6.7.

Equation (6.28) can be rearranged, noting that $x_1+x_2 = 1$, to the form

$$x_1^m = \frac{Kx_1^l}{1+(K-1)x_1^l} \tag{6.29}$$

which is an individual isotherm. It will be noted that the form of (6.29) closely resembles that of the Langmuir equation (5.17) for the adsorption of a single gas on a solid.

Equation (6.28) will be of only limited use in practice owing to the simplicity of the assumed model. However, theories have been developed for the adsorption from regular solutions and from mixtures of molecules of different sizes. In addition Ash *et al.* (1968) have considered multilayer adsorption from monomer+dimer mixtures. At present, however, there are insufficient data available with which to carry out detailed tests of the various theories of adsorption at the solid–liquid interface.

6.9. Adsorption from dilute solutions: introduction. The adsorption on to solids from dilute solution is of great practical interest. It is of importance in such processes as dyeing, detergence, purification of liquids, lubrication, chromatography etc. A detailed theoretical treatment is, however, more complicated than for adsorption at the liquid–vapour interface, largely as a result of the ill-defined, heterogeneous nature of the surfaces of many solids. Nonetheless, a great deal of experimental work has been reported.

As for liquid mixtures, the adsorption is usually determined by the measurement of the change in the composition of a solution which is allowed to attain equilibrium with a known mass of solid. For a sufficiently dilute solution of solute 2 in solvent 1, $x_1 \simeq 1$ and $x_2 \simeq 0$ and it follows from (6.14) that

$$\frac{n_0 \Delta x_2}{m_A} \simeq n_2^s. \qquad \blacktriangleleft\textbf{(6.30)}$$

The concentration Γ_2^s of the solute is related to n_2^s by

$$\frac{n_2^s}{\Sigma} = \Gamma_2^s, \tag{6.31}$$

where Σ is the specific surface area of the adsorbent. The subscript 2 will be omitted in future except where parameters for both solvent and solute occur together. It should be stressed that (6.30) does not imply that n_1^s is zero but only that $n_1^s x_2$ is very small. The approximation (6.30) becomes less valid as the concentration of the solution increases, and care should be exercised before $n_0 \Delta x / m_A$ is taken as representing the solute adsorption alone.

Adsorption isotherms (plots of n^s or Γ^s against x) of widely varying shapes have been observed, as might be expected. Giles *et al.* (1960) have given a classification of the isotherms for the adsorption of dissolved solids, which consists of no less than eighteen types. Probably the most commonly observed form, however, is that which closely resembles the isotherm of type I for the adsorption of gases on solids (§5.7). For this reason, data have often been described in terms of the Langmuir equation (§1.12 and §5.9). The latter has been derived (§1.12) for an ideal localised monolayer of a single component. This model is clearly inappropriate for films at the solid–solution interface since, amongst other possible objections, the films contain solvent as well as adsorbate. Nevertheless, an isotherm with the same form as the Langmuir equation can be derived for monolayer adsorption from dilute solution in the following way. Suppose that the solid surface is homogeneous and that the solvent (1) and the solute (2) molecules have equal molar surface areas, a. Suppose, further, that both the bulk and surface phases are ideal. Under these circumstances, at adsorption equilibrium, the equation

$$\gamma = \gamma_1^0 + \frac{RT}{a} \ln \frac{x_1^m}{x_1^l} \tag{3.35}$$

should be applicable, where γ refers to the solid–solution interface and γ_1^0 to the interface between the adsorbent and pure solvent. Noting that π, the surface pressure, is $\gamma_1^0 - \gamma$, and that for a dilute solution, $x_1^l \simeq 1$, (3.35) may be written

$$\pi = -\frac{RT}{a} \ln x_1^m. \tag{6.32}$$

Since the adsorbed film is assumed to be monomolecular it follows from (1.47) that

$$\Gamma_1^s + \Gamma_2^s = 1/a. \tag{6.33}$$

In addition, $x_1^m = \Gamma_1^s/(\Gamma_1^s + \Gamma_2^s)$ so that, using (6.32) and (6.33) it follows that

$$\pi = -\frac{RT}{a} \ln \left(\frac{N_A \dot{a} - a}{N_A \dot{a}} \right), \tag{6.34}$$

where Γ_2^s has been written as $1/N_A\dot{a}$. Comparison of (6.34) and (1.96) shows that the former corresponds to an isotherm of the same form as the Langmuir equation. It will be appreciated, however, that the assumptions made in the derivation of (6.34) are not likely to be valid for many systems.

6.10. Determination of the specific surface areas of solids by adsorption from solution. The isotherms for adsorption from solution, like gas–solid isotherms, have often been used to determine the specific surface areas of solids. Thus, for example, if the adsorption from solution yields an isotherm which can be fitted by the Langmuir equation, the saturation adsorption of the solute (equivalent to v_m in gas adsorption) can be obtained. Then, if the cross-sectional area of a solute molecule at the surface is known, the specific surface area of the solid can be calculated. The method has the advantage over gas adsorption that no complex gas handling apparatus (see fig. 5.1) is required. If the solute used is coloured (e.g. a dye such as methylene blue) the change in concentration of the solution, and hence the adsorption, can be determined colorimetrically. For colourless solutions the change in the refractive index of the solution may be measured.

A serious limitation of the method lies in the fact that the cross-sectional areas of the adsorbate molecules are not usually known with certainty. The more commonly used and strongly adsorbed solutes (e.g. long chain fatty acids and alcohols etc.) are asymmetrical, and the appropriate cross-sectional area depends on the orientation of the adsorbate at the interface. The area of, say, n-octadecanol in the vertical orientation is about 0.2 nm^2 $molecule^{-1}$, whereas in the flat orientation the area may be of the order of 1.0 nm^2 $molecule^{-1}$. Furthermore, the orientation assumed by a particular solute may depend upon the solvent and the adsorbent used. Nevertheless, the method is useful for the rapid determination of relative values of the surface areas of different samples of the same material.

6.11. Thermodynamic parameters of adsorption. The thermodynamic parameters of adsorption are used to express the equilibrium adsorption properties of a given system in numerical terms, and are useful in the understanding of the nature of the forces responsible for adsorption. For example, if one wished to describe numerically the adsorption characteristics of a system comprising a fibre in contact with a solution of a dye, the standard free energy of adsorption of the dye is a useful quantity. Its value is independent of the concentration of the solution, the amount of fibre present and the extent of adsorption. The characteristics of different systems can therefore be compared in terms of the appropriate values of the standard free energies of adsorption.

For an infinitely dilute film of component 2 adsorbed from solvent 1, the chemical potential μ of the adsorbed solute, may be written

$$\mu = \mu^{\ominus, s, +} + RT \ln \Gamma^s. \tag{6.35}$$

The expression (6.35) can be obtained in a similar manner to (3.76). Since, for the infinitely dilute film, $\pi = \Gamma^s RT$

$$\mu^{\ominus, s, +} = \mu^{\ominus, s'} + RT \ln RT. \tag{6.36}$$

It then follows that the standard free energy of adsorption $\Delta_a \mu^{\ominus, +}$ is given by

$$\Delta_a \mu^{\ominus, +} = -RT \ln \frac{\Gamma^s}{x} \tag{6.37}$$

and differs from $\Delta_a \mu^{\ominus'}$ of (3.78) by $RT \ln RT$. The standard states in (6.37) are $\Gamma^s = 1$ and $x = 1$ for the surface and bulk respectively. Since (6.37) is for ideal systems, Γ^s/x is the limiting slope, at low x, of the adsorption isotherm.†

It is often difficult to determine the adsorption on solids from dilute solutions with great accuracy. For this reason it may not be possible to obtain the limiting slope of a Γ^s against x curve with the desired precision. It has been mentioned (§6.9), however, that the isotherms frequently have a Langmuirian form, in which case a standard free energy of adsorption may instead be obtained from (1.95), i.e.

$$\Delta_a \mu^{\ominus} = -RT \ln \left(\frac{\theta}{1-\theta} \bigg/ x \right), \tag{6.38}$$

where the standard states are $\theta = \frac{1}{2}$ and $x = 1$. It will be appreciated that, to calculate θ or Γ^s, Σ must be known.

A standard enthalpy of adsorption, $\Delta_a \mathrm{H}^{\ominus}$ can be obtained from $\Delta_a \mu^{\ominus}$ using the well-known expression

$$\Delta_a \mathrm{H}^{\ominus} = \frac{\partial(\Delta_a \mu^{\ominus}/T)}{\partial(1/T)}. \tag{6.39}$$

† Some authors have expressed surface concentrations in bulk units. This necessitates the division of Γ^s by the film thickness. The choice of film thickness must, to some extent, be arbitrary. Although, in certain circumstances, this procedure may be useful it is by no means necessary.

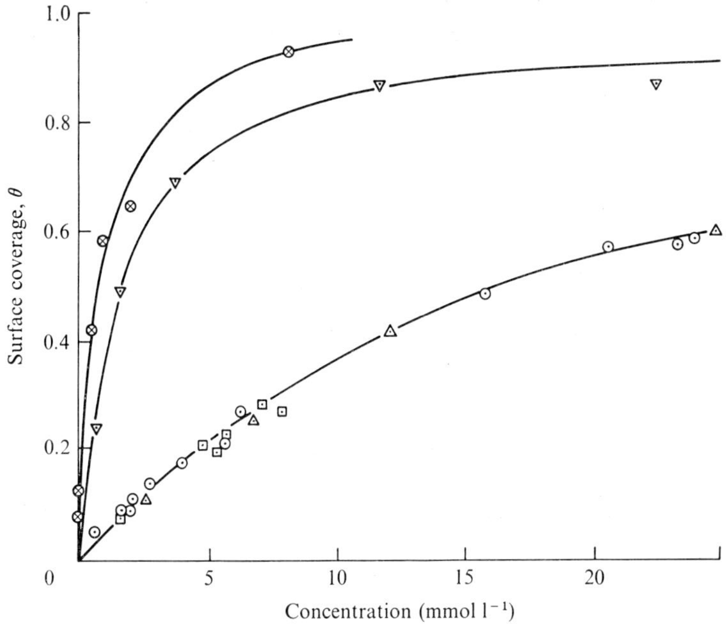

Fig. 6.9. Isotherm for the adsorption of alcohols from solution in benzene on to alumina. ⊙, octadecanol; △, hexadecanol; ⊡, triacontanol; ▽ cholesterol; ⊗, phenol. (From Crisp, 1956.)

Although the values of the standard free energies of adsorption obtained from (6.37) and (6.38) will usually be different, the standard enthalpies obtained from these free energies will be the same since the standard states for both free energies have been chosen in terms of surface coverage. The standard enthalpy derived from $\Delta_a\mu^{\ominus\prime}$ (3.78), however, is different by an amount RT.

Some of the procedures discussed in §6.10 and the present section are well illustrated by a study of the adsorption of various alcohols and phenol from dilute solutions in benzene on to alumina (Crisp, 1956). The adsorption isotherms (fig. 6.9) could all be fitted by the Langmuir equation. It was found that, for the n-alkanols, the adsorption was independent of the chain length. It was assumed that the alkanols were oriented vertically at the alumina surface, and the cross-sectional areas of all three alkanols were taken as 0.205 nm^2 molecule^{-1}. The saturation adsorption was obtained from a plot according to a linear form of the Langmuir equation (see 5.20); in this way a value of Σ for the alumina of 91 $m^2 g^{-1}$ was obtained. This was in good agreement with the value of 96 $m^2 g^{-1}$ derived from the adsorption of cholesterol, where the cross-sectional area was taken as 0.38 nm^2 $molecule^{-1}$, which is the value obtained for a spread film at the air–water interface.

Values of the standard free energy of adsorption obtained by means of an expression similar to (6.38) but containing a surface thickness, were the same for the three alkanols. Values of the standard heats of adsorption of octadecyl (C_{18}) and melissyl (C_{31}) alcohols were, respectively, 30.7 and 31.5 kJ mol^{-1}. Since the heats and free energies are essentially independent of the alkanol chain length, it was concluded that the adsorption results mainly from the interaction of the polar groups with the surface of the alumina. In addition, since the Langmuir equation holds over a wide range of coverage, it was supposed that the energy of the adsorption sites was uniform and that the adsorption was localised (see §1.12). It should be pointed out, however, that the Langmuir equation can be derived without making the assumption of localisation (§6.9). It follows, therefore, that obedience to the Langmuir equation need not necessarily imply localisation.

6.12. π against $\dot{a}$ curves for adsorbed films at the solid–liquid interface. Much consideration has been given in the past to obtaining appropriate equations of state for films adsorbed at liquid–vapour, liquid–liquid and solid–gas interfaces. Less attention appears to have been given to the solid–liquid interface, but π against $\dot{a}$ curves can be obtained from adsorption data.

The Gibbs equation for adsorption of a single solute from a sufficiently dilute solution on to an insoluble solid may be written (see p. 103)

$$\Gamma^s = -\frac{1}{RT}\frac{d\gamma}{d\ln x}, \tag{6.40}$$

where γ refers to the solid–liquid interface. Noting that $d\pi = -d\gamma$ integration of (6.40) yields the expression for π

$$\pi = RT\int_{x=0}^{x}\Gamma^s d\ln x \tag{6.41}$$

which is analogous to (5.13) for the solid–gas interface. It follows that π may be calculated from values of Γ^s at various values of x.

Plots of π against $\dot{a}$, obtained using (6.41), have been reported by Daniel (1951) in connection with an investigation into problems asso-

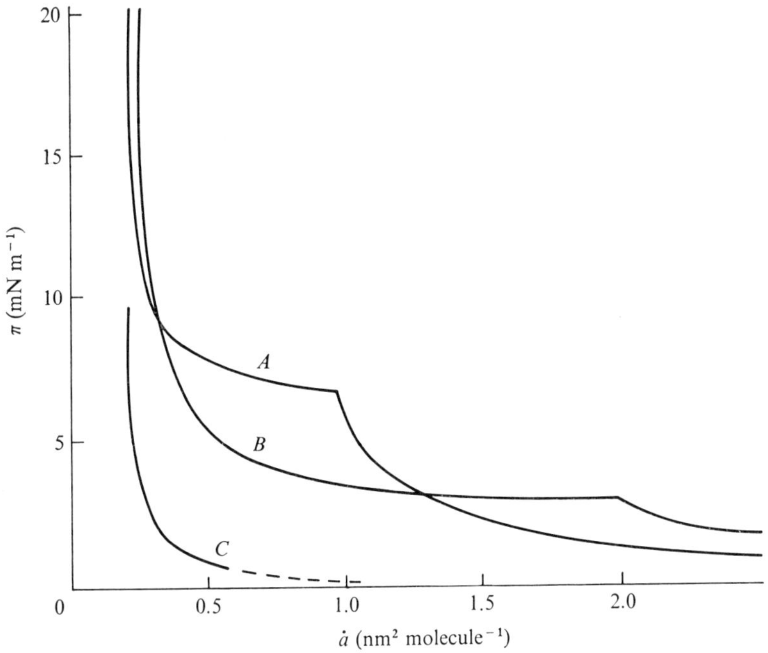

Fig. 6.10. π versus $\dot{a}$ curves for octadecanol (*A*), ethyl octadecanoate (*B*), and octadecanoic acid (*C*) at the benzene–nickel interface. (After Daniel, 1951.)

ciated with lubrication. The results for various films adsorbed at the interface between nickel and benzene solution are shown in fig. 6.10. Although accurate values of π could not be obtained, the general shapes of the curves are interesting because there is a marked similarity between these curves and those obtained for films at the liquid–vapour interface.

6.13. The adsorption of polymers from solution. The adsorption of polymers from solution has been widely studied and may reasonably be regarded as a special branch of surface chemistry. The theoretical treatment of polymer adsorption is complex and it is not intended here to discuss the various surface equations of state and adsorption isotherms which have been proposed. Rather, an indication is given of the scope of the topic and of the special considerations involved in the study of polymer adsorption.

Early experimental studies were performed using biological polymers (e.g. proteins) but more recently many synthetic polymers have been investigated as a consequence, no doubt, of their great practical importance.

Various experimental difficulties exist in the study of the adsorption of polymers from dilute solution. One problem is that a system may take several days or even weeks to attain equilibrium. Furthermore, many polymers are polydisperse, i.e. there is a distribution of molecular weights within the sample. Often, appreciable polydispersity persists even after fractionation.

The experimental facts concerning the adsorption appear to be reasonably well established so that it is clear what an adequate theory must explain. The amount of a polymer adsorbed (on a fixed area) rises very rapidly with the bulk concentration and reaches a plateau (see fig. 6.11). In some instances the rise is so steep that this region of the isotherm is not experimentally accessible. The amount of polymer adsorbed increases with molecular weight at low molecular weights. At high molecular weights, however, no such dependence can be detected. The temperature is found to have only a small influence on the extent of adsorption and, as might be expected, the adsorption is greater from the poorer solvents. Such an effect is seen in fig. 6.11 for the adsorption of a polyvinyl butyral polymer (XYSG) from various solvents on iron powder.

Several theoretical treatments of polymer adsorption have been advanced, notable amongst which are those of Simha, Frisch & Eirich (1953) and Silberberg (1968). One of the main considerations in any theory is that of the configurations of the polymer at the interface. Unlike small adsorbate molecules, polymers have many groups which are capable of attachment to the adsorbent surface. The segments between the adsorbed groups can thus form loops which extend into the solution. The two theories mentioned above, however, differ in the number of points of attachment and the loop sizes which are predicted.

The comparison of experimental with theoretical isotherms is not a sensitive test of the validity of a model. This has been mentioned elsewhere in connection with other systems. In the present instance it is more likely that support for a model will come from the experimental determination of relevant parameters, such as the number of polymer-adsorbent contacts and the dimensions of adsorbed layers. For example, viscosity measurements may give an indication of film thickness. Infra-red spectra of molecules at surfaces may indicate the number of free and attached groups. An example of such an investigation is reported by Fontana & Thomas (1961). The number of attached seg-

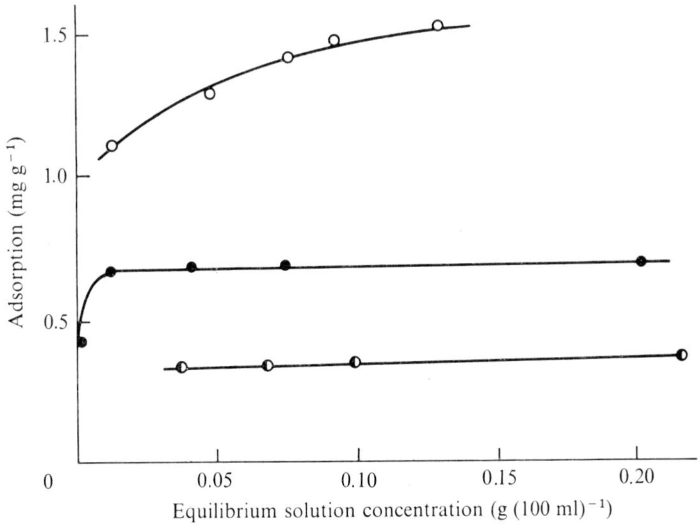

Fig. 6.11. Adsorption isotherms for polymer XYSG adsorbed on iron at 30.4 °C from carbon tetrachloride (○), benzene (●), and chloroform (◐). (After Koral, Ullman & Eirich, 1958.)

ments of poly-(alkyl methacrylate) molecules on silica was determined. The attachment takes place by H-bonding between the carbonyl oxygen of the ester group and the –OH groups on the surface. The characteristic stretching vibration of the carbonyl group is shifted to lower frequencies when hydrogen bonding occurs. In this way, it was found that about-one third of the segments were attached to the silica. This fraction was apparently independent of the molecular weight of the polymer.

Other useful techniques for investigating the nature of adsorbed polymers are discussed by Rosoff (1969) in a review of the surface chemistry of polymers.

6.14. The adsorption of ions. The basic theory of ion adsorption has been given in chapter 2 and the special cases of adsorption at liquid interfaces and at polarised and non-polarised electrodes have been discussed in chapters 3 and 4 respectively. The only important difference from the theory of the adsorption of non-electrolytes is that the standard free energy of adsorption is a function of the electrical potential difference between the solid surface and the bulk of the solution. Thus, instead of the $\Delta_a\mu^\ominus$ of (6.38), the standard electrochemical potential of adsorption $\Delta_a\tilde{\mu}^\ominus_{+,-}$, defined in chapter 2 must be used, where

$$\Delta_a\tilde{\mu}^\ominus_{+,-} = \Delta_a\mu^\ominus_{+,-} + z_{+,-}e^-(\varphi(\delta) - \varphi(\infty)). \tag{2.43}$$

$z_{+,-}$ is the (algebraic) valence of the ion in question, $\varphi(\delta)$ is the electrical potential in the plane of the adsorbed ions and $\varphi(\infty)$ is the potential in the bulk of the solution. As in §3.8, $\varphi(\infty)$ may, for present purposes, be taken as zero. With the exception that the surface potential is written differently, the right-hand side of (2.43) is clearly recognisable in (3.95) for the adsorption of ions at liquid interfaces, and is also seen in (2.45), which is generally referred to as the Stern equation. For convenience (2.45) is restated, i.e.

$$N_1 = \frac{N_s x \exp[-\Delta_a\mu^\ominus/kT] \exp[ze^-\varphi(\delta)/kT]}{1 + x \exp[-\Delta_a\mu^\ominus/kT] \exp[ze^-\varphi(\delta)/kT]}, \quad \blacktriangleleft\textbf{(6.42)}$$

where N_1 is the number of ions adsorbed and N_s is the total number of adsorption sites per unit area of surface, x and z are the mole fraction and valence of the ion in question and $\varphi(\infty)$ has been taken as zero.

Equation (6.42) is Langmuirian in form and, apart from the electrostatic terms, resembles (6.38). The adsorption of ions at solid surfaces is usually assumed to follow this type of equation more closely than that for non-localised adsorption (3.95). This assumption may at first sight seem somewhat arbitrary but it is not unreasonable to suppose that ion adsorption on to many solids takes place preferentially at ionic, or at least partially ionic, lattice points.

As the free energy of adsorption of ions involves both short range (chemical) and long range (electrical) terms it is a much more complicated quantity than that for non-electrolytes. Unlike $\Delta_a\mu^\ominus$ (cf. (6.38)), $\Delta_a\tilde{\mu}^\ominus_{+,-}$ will not be a constant unless the term containing $\varphi(\delta)$ is small compared to $\Delta_a\mu^\ominus_{+,-}$. This situation may arise when the ion has a large non-ionic group attached to it (e.g. an alkyl chain) which interacts strongly with the interface so giving a large $\Delta_a\mu^\ominus_{+,-}$. A further possibility is that instead of, or in addition to, a large $\Delta_a\mu^\ominus_{+,-}$, $\varphi(\delta)$ may be suppressed by the presence of a high concentration of non-adsorbing electrolyte (see §2.6). Usually, however, the term in $\varphi(\delta)$ is not insignificant and, moreover, varies as the adsorption of ions varies. The effect of this, as pointed out in §3.8, is to reduce progressively the magnitude of the free energy of adsorption as the adsorption increases.

Owing to the presence of the electrical potential, the testing of an equation such as (6.42) is much more difficult than for a comparable equation for non-electrolytes. The problems of estimating $\varphi(\delta)$ and other surface potentials, have been stressed in chapter 2 and a variety of approaches have been used by different investigators. Probably the most common and useful is that which assumes that $\varphi(\delta)$ may be equated to the electrokinetic potential. The evidence that this procedure is permissible is not wholly convincing, and the assumption may be much better in some systems than others. Nevertheless, the measurement of electrophoretic mobility or streaming potential (§2.8) is usually fairly easy, and the picture of the adsorption process so obtained is frequently very helpful. This general approach is well described by Ottewill, Rastogi & Watanabe (1960) and was used by these authors in subsequent experimental work on the adsorption of ionic surface active agents on to silver iodide sols (Ottewill & Watanabe, 1960).

The measurement of ion adsorption by solids presents no more difficulty than does the adsorption of non-electrolytes. The presence of a large surface area is obviously an advantage and hence most investigations have been carried out with suspensions of the finely divided solid. For systems of low surface area, but which have appreciable flat surfaces (e.g. single crystals), methods which involve radio-isotopes have been developed.

For obvious practical reasons many investigations of ion adsorption have been carried out with silver halides. Some description of this work has been given in chapter 4. The adsorption of ions by minerals is, however, of great practical importance (e.g. in connection with soil science) and has been relatively well studied. The methods employed have usually been variations on those outlined above. For solids such as silica, which are essentially gels and may be porous to ions, it is found that the surface charge density calculated, for example, from a BET surface area and the number of ions taken up, may apparently rise indefinitely with the concentration of the ion in the bulk phase. The electrokinetic potential may not, however, exhibit similar behaviour and, indeed, may not be at all high. This result suggests that the adsorbing ion may penetrate well into the matrix of the gel and that a simultaneous penetration of counter-ions may neutralise the space charge so created. This particular situation is well illustrated by the results of Tadros & Lyklema (1968) for the adsorption of hydroxyl ions on to silica (fig. 6.12). The penetration of adsorbing ions beyond the geometrical surface of a particle is very common for biological systems. Thus, although the incidence of surfaces or interfaces in biological specimens is very high, the measurement of adsorption at these inter-

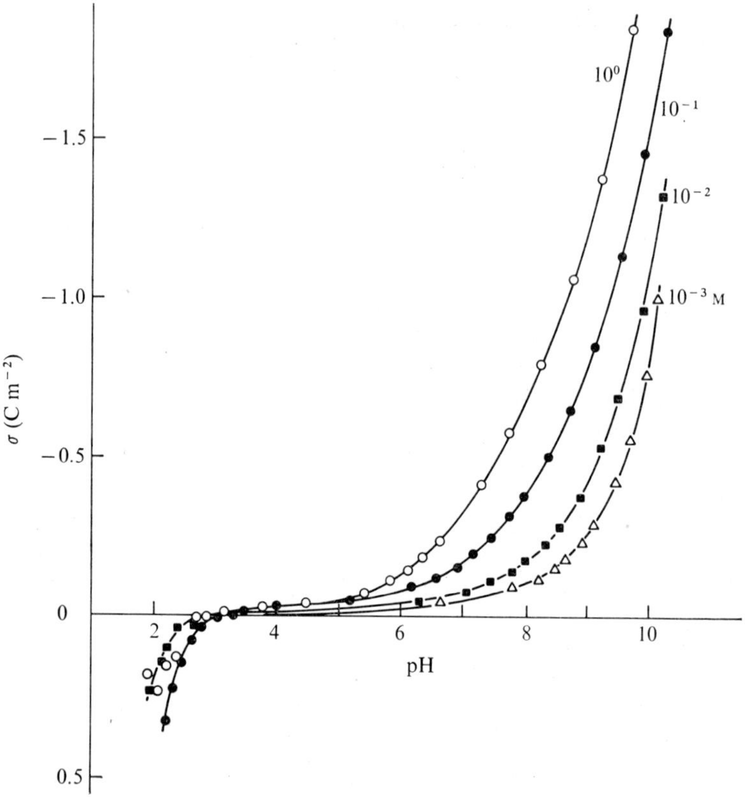

Fig. 6.12. The surface charge density of a porous silica as a function of pH in various potassium chloride solutions. (After Tadros & Lyklema, 1968.)

faces it not a simple matter as they are mostly gel-like in character. Direct estimates of the uptake of ions are, as for porous silica, often far in excess of those indicated by electrokinetic measurements and, for this reason, the latter technique has been widely employed in the study of the surfaces of biological cells and other small particles.

References

Adam, N. K. & Elliott, G. E. P. (1962). *J. Chem. Soc.*, 2206.
Ash, S. G., Everett, D. H. & Findenegg, G. H. (1968). *Trans. Faraday Soc.* **64**, 2645.
Corkill, J. M., Goodman, J. F. & Tate, J. R. (1966). *Trans. Faraday Soc.* **62**, 979.
Crisp, D. J. (1956). *J. Colloid Sci.* **11**, 356.
Daniel, S. G. (1951). *Trans. Faraday Soc.* **47**, 1345.
Ellison, A. H., Fox, H. W. & Zisman, W. A. (1953). *J. Phys. Chem.* **57**, 622.

Everett, D. H. (1964). *Trans. Faraday Soc.* **60**, 1803.
Everett, D. H. (1965). *Trans. Faraday Soc.* **61**, 2478.
Fontana, B. J. & Thomas, J. R. (1961). *J. Phys. Chem.* **65**, 48.
Gasser, C. G. & Kipling, J. J. (1960). *J. Phys. Chem.* **64**, 710.
Giles, C. H., MacEwan, T. H., Nakhwa, S. N. & Smith, D. (1960). *J. Chem. Soc.*, 3973.
Gould, R. F. (editor) (1964). *Contact Angle, Wettability and Adhesion* (ACS, Washington).
Kipling, J. J. (1965). *Adsorption from Solutions of Non-Electrolytes* (Academic Press): (*a*) pp. 172–5; (*b*) pp. 32 *et seq.*
Koral, J., Ullman, R. & Eirich, F. R. (1958). *J. Phys. Chem.* **62**, 541.
Langman, C. A. J. (1965). Ph.D. thesis, Hull University.
Minkoff, G. J. & Duffett, R. H. E. (1964). *B. P. Magazine* **13**, 17.
Nagy, L. Gy. (1963). *Periodica Polytechnica*, **7**(2), 75.
Ottewill, R. H., Rastogi, M. C. & Watanabe, A. (1960). *Trans. Faraday Soc.* **56**, 854.
Ottewill, R. H. & Watanabe, A. (1960). *Kolloid-Z.* **170**, 132.
Rosoff, M. (1969). *Physical Methods of Macromolecular Chemistry* (ed. B. Carrol), (Dekker): chapter 1.
Schay, G. & Nagy, L. Gy. (1961). *J. chim. phys.*, 149.
Schay, G., Nagy, L. Gy. & Szekrenyesy, T. (1962). *Periodica Polytechnica*, **6**, 91.
Silberberg, A. (1968). *J. Chem. Phys.* **48**, 2835.
Simha, R., Frisch, H. L. & Eirich, F. R. (1953). *J. Phys. Chem.* **57**, 584.
Tadros, Th. F. & Lyklema, J. (1968). *J. Electroanal. Chem.* **17**, 267.
Zisman, W. A. (1964). *Contact Angle, Wettability and Adhesion* (ed. R. F. Gould), (ACS, Washington): chapter 1.

Index